FROM MARGINAL GAINS TO A
CIRCULAR REVOLUTION

FROM MARGINAL GAINS TO A
CIRCULAR REVOLUTION

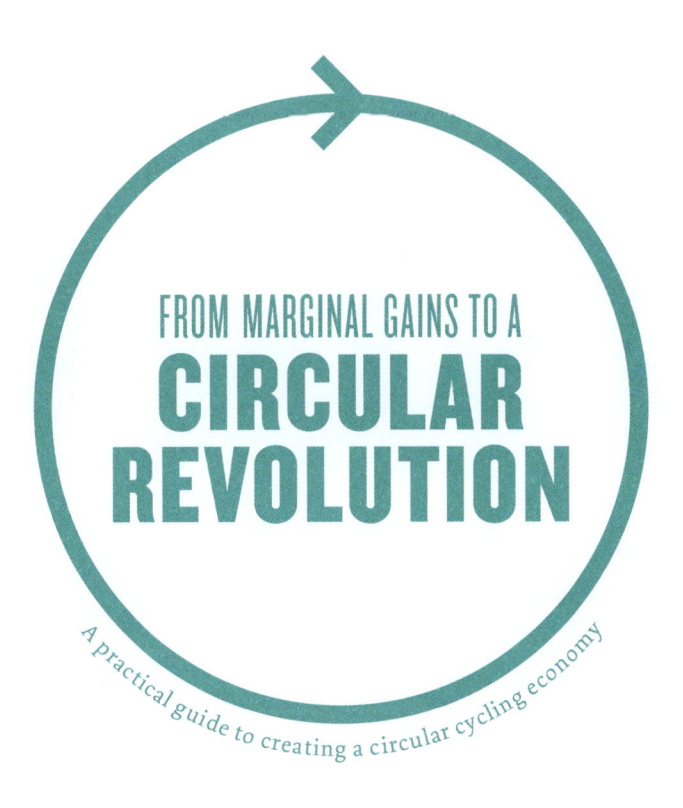

A practical guide to creating a circular cycling economy

ERIK BRONSVOORT & MATTHIJS GERRITS

Warden Press

ISBN:
Paperback (full colour): 978-94-92004-93-2
Paperback (black & white): 978-94-92004-94-9
E-book: 978-94-92004-95-6

Cover design, interior design and lay-out: Pankra.com & Anihow
Photo authors: © the authors

© 2020 Erik Bronsvoort, Matthijs Gerrits

Published by Warden Press, Amsterdam

All rights reserved. No part of this publication may be reproduced, stored in a retrieval system, or transmitted in any form or by any means, electronic, mechanical, photocopying, recording or otherwise without the prior written permission of the publisher.

wardenpress.com

FOR
TARA & LUUK
AND
SAAR

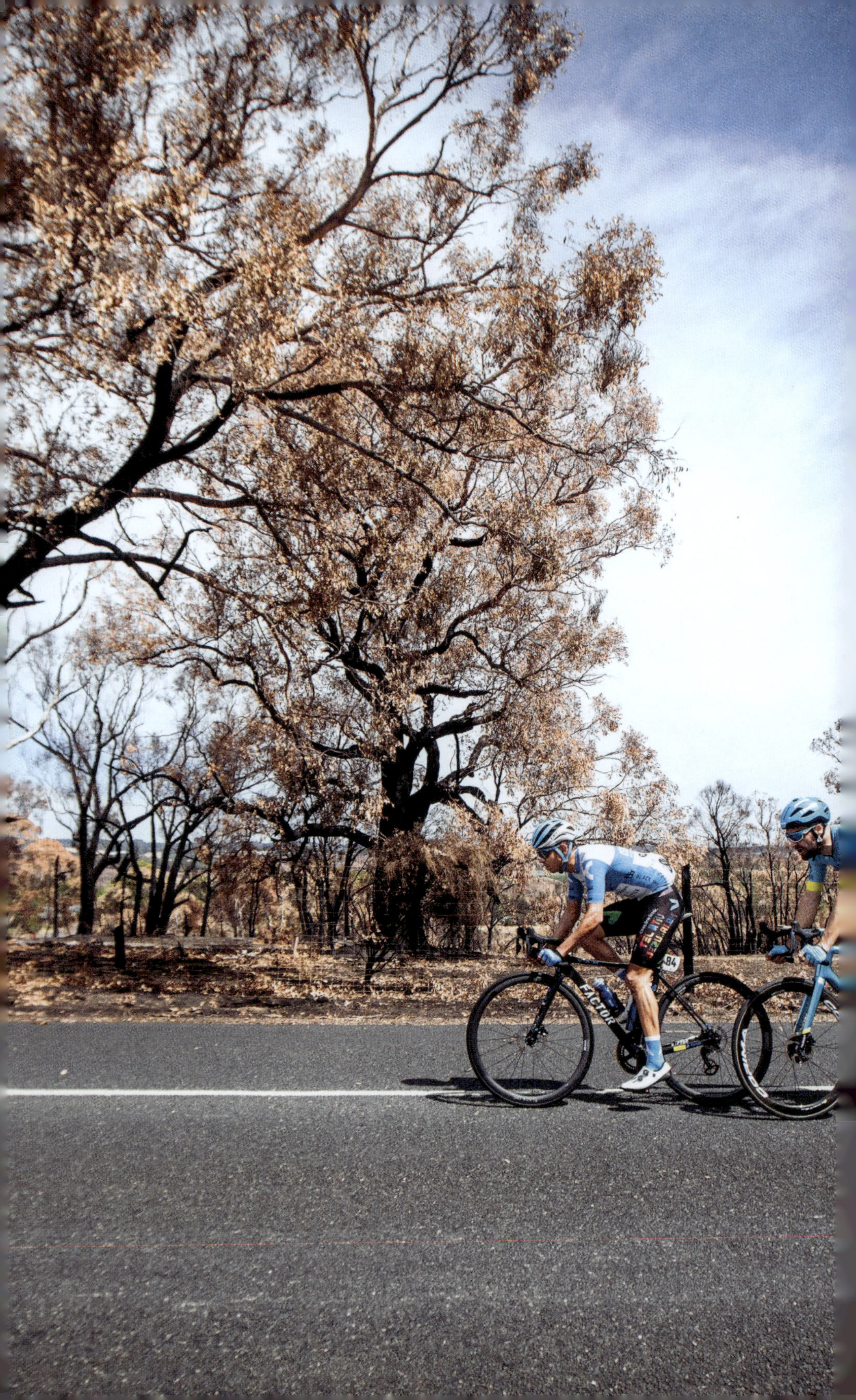
FACTOR

CONTENTS

PROLOGUE		9
THE ROADBOOK TO A REVOLUTION		17

STAGE 1 THE COMPLEX WORLD OF CYCLING AND SUSTAINABILITY — 23
- 1.1 Bike design, rules & regulations — 24
- 1.2 The tragedy of the commons — 33
- 1.3 Governing the cycling industry — 34

STAGE 2 THE LINEAR CYCLING ECONOMY — 39
- 2.1 The linear economy — 41
- 2.2 Linear design incentives — 43
- 2.3 Limits to growth — 45
- 2.4 Measuring environmental impact — 51

STAGE 3 CREATING VALUE IN A CIRCULAR CYCLING ECONOMY — 55
- 3.1 From linear to circular – where are the opportunities? — 58
- 3.2 Circular business models — 78

STAGE 4 ACTION PLAN FOR A CIRCULAR CYCLING INDUSTRY — 87
- 4.1 The business model of the circular bike — 91
- 4.2 Sarah's circular bike experience — 109
- 4.3 Circular design incentives — 112

STAGE 5 GETTING THERE — 117
- 5.1 Managing the transition — 118
- 5.2 Contribution of each stakeholder — 124

FINISH STAND UP FOR A REVOLUTION — 139

THANKS	143
ABOUT THE AUTHORS	144
NOTES	152

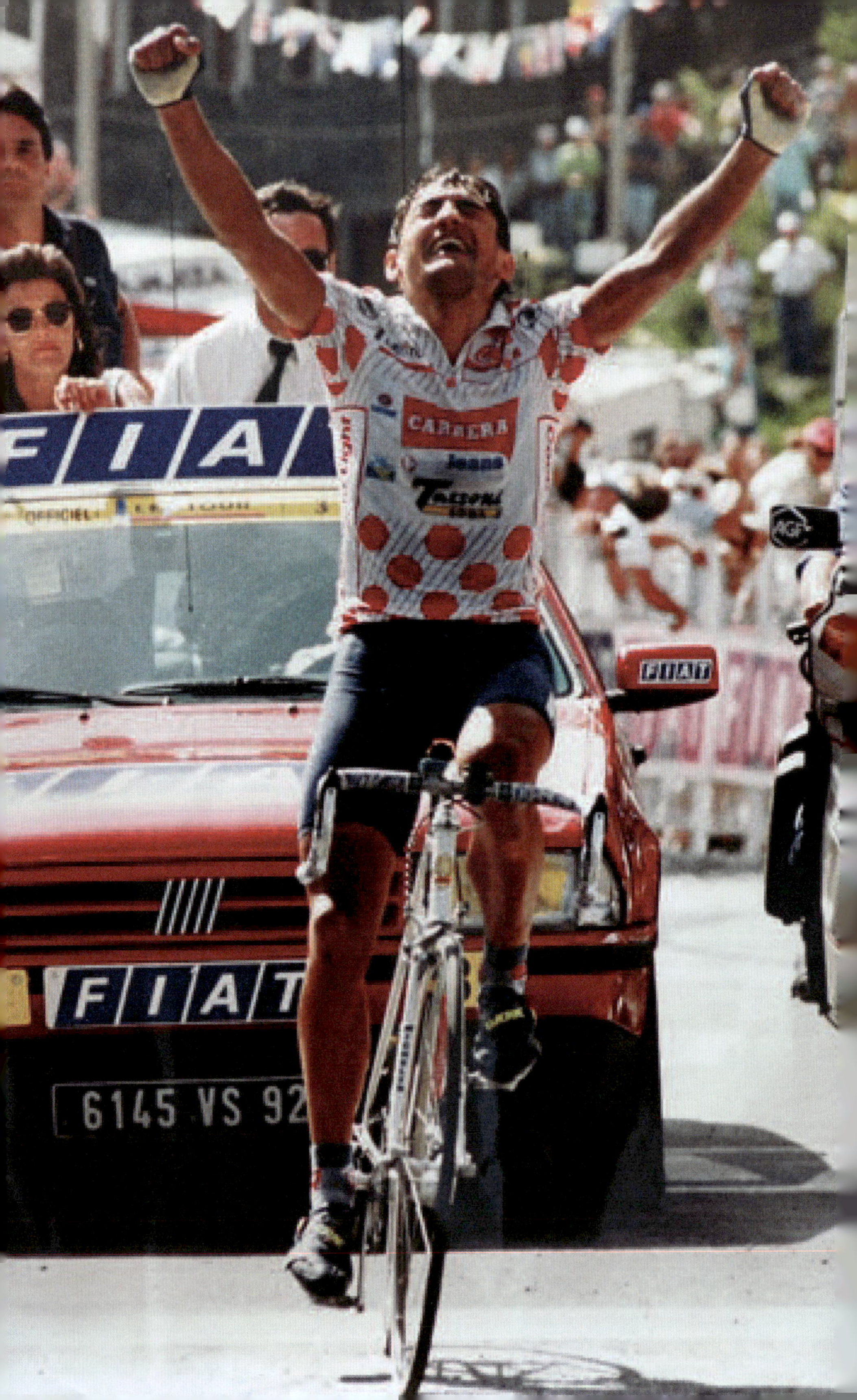
FIAT
CARRERA
Jeans
FIAT
FIAT
6145 VS 92

PROLOGUE

On a beautiful morning in July a couple of decades ago, the heat of the French summer warms the legs of the riders awaiting the start of the first mountain stage in the Alps. It is the morning of the queen stage of the Tour the France that year.

The fatigue in the legs of the riders has built up in the days leading up to this daunting 254km stage from St Gervais to Sestriere. It is one of the first days since the start of the Tour de France almost two weeks ago that the sun is out. The riders have had an exceptionally cold and rainy Tour so far, and to make things worse, the pace has been extremely high in stages that felt longer than normal. Both riders and staff are suffering.

Many complain about the enormous distances that need to be covered, not just during the stages but also in the transfers between them. This year's Tour takes the riders through no fewer than seven European countries in three weeks: 3,983km in 21 stages. Spain, the first country, hosted the prologue and a couple of stages. Next up was a team time trial and another couple of stages in France, all before the riders rode from Roubaix to Brussels over wet and slippery cobbles. Then from Belgium to the Netherlands and onwards into Germany.

The individual time trial of over 65km in Luxembourg resulted in the first serious time gaps in the general classification, but still none of the favourites is wearing the yellow jersey. New names in the peloton challenge the older generation with relentless attacks; the older generation retaliates with the same strategy. Only one man keeps his cool, as he has done over the many stage races he won in the years leading up to this Tour.

Now, after twelve stages, it is time for the first proper mountain stage. With 254km and five climbs (one of the 2nd category, three of the 1st and one hors catégorie) it's a good thing the sun is out. Exactly 40 years before, the legendary Fausto Coppi won this same stage. That victory inspired a young Italian rider. He attacks on the first climb for a monster solo of over 245km. The young Italian is Claudio Chiappucci. He has been wearing the polka dot jersey for the best climber since the day he earned enough points to take the jersey in stage 9. He took it off another young rider's shoulders, the Frenchman Richard Virenque. Virenque is on his debut Tour, got the jersey after the second stage, and is already destined to become a French national hero.

Chiappucci increases his lead in the polka dot ranking to an enormous gap by taking full points on every single climb of the day. The crowd lining the final climb to Sestriere goes crazy, running alongside Chiappucci. With just over a kilometre to go, Chiappucci has to push motorbikes and spectators to get them out of his way and reach the top, hanging on to his ever-decreasing lead over Miguel Indurain – the unbeatable star rider of this era. His unbelievable 245km solo delivers him the most impressive win of his career – and a second place in the general classification, behind the new leader, Miguel Indurain.

The next morning no fewer than eighteen riders abandon the race, with Indurain wearing the yellow jersey again for the first time since the prologue. He would take the yellow jersey to Paris in what would be the second of five consecutive Tour de France wins for the Spanish rider.

The year was 1992, a time when few riders were wearing helmets and carbon fibre frames were a novelty, slowly replacing the steel framesets that had been part of the peloton's gear for so many decades. Shifters integrated into the brake levers were a recent innovation – Shimano had introduced the Dura Ace 7400 Dual Control Shifters in 1990 and Campagnolo launched the Record 8 Speed Ergopower in late 1991. For the mountain stages, some riders still preferred the levers mounted on the downtube of the frameset to save a few grams.

The queen stage of the 79th edition of the Tour de France took place a little over a month after 154 nations signed the UNFCCC – the United Nations Framework Convention on Climate Change at the 1992 Rio de Janeiro Earth Summit. The Cold War had ended the previous year with the collapse of the Soviet Union. The approximately 5.5 billion people living on planet Earth were no longer divided by an East-West conflict. It felt as if the world was coming together again, embarking on a new and peaceful future with global opportunities resulting in lower poverty levels all over the world, as well as a better environment.

World leaders like the US President George H.W. Bush, UK Prime Minister John Major and Cuba's leader Fidel Castro were all present at the Earth Summit and made promises that they would act on issues such as toxic waste, air pollution, water use and fossil fuels. For the first time in history, a vast majority of countries agreed that it was required to 'stabilize the emissions of greenhouse gas concentrations in the atmosphere at a level that would prevent dangerous anthropogenic interference with the climate system.'[1] In everyday language: it was necessary to make sure greenhouse gas emissions caused by humans would not lead to climate change beyond a safe level.

Since 1992, when Miguel Indurain won his second Tour de France, a lot has changed – generally for the better. The number of people living in poverty has dropped from some 34% to less than 10% of the world population,[2] even though the latter surged to approximately 7.8 billion people by 2020.[3] With the increase of population and wealth, the market for bicycles grew as well, meaning that more people enjoy riding a bike than ever before.

However, there is also a downside. Population growth and increased wealth have taken their toll on the environment. We are living in what we call a linear economic system, where we use finite sources of materials and fossil fuels to make products and send them all over the world to consumers. We often use these products only briefly before they end up in a landfill, in an incinerator or, worse, in the soil or the oceans. The production processes, transportation and use of the products cause emissions and pollution. In a linear economy, We Take, Make and Waste.

Because of the growth in population and wealth, our factories use ever more resources to produce ever more goods, generating ever more waste. Air quality is poor in many of the world's cities. We consume more energy than at any time before to heat and cool our buildings and to move goods and people across the planet. Global CO_2 emissions have not stabilised since 1992; instead, they have continued to rise. As a result, average temperatures in the 2010s were about 0.5°C above those of the 1980s. Extreme weather conditions and natural disasters have increased in frequency and intensity, just as forest fires have.[4]

The 1992 UNFCCC can be seen as a starting point for change, which has been slow to gather momentum. This is partly because it has taken time to produce sufficient evidence that climate change is really happening, required to convince politicians, business and citizens that they need to act. Citizens now use their votes and their wallets to demand change. And more and more companies see they need to change to avoid a backlash from regulators, to save costs and to gain a competitive advantage in a market with new customer demands.

Electricity produced from fossil fuels is increasingly being replaced by electricity from renewable sources such as the sun and wind. Sales of electric cars show double-digit growth in a number of countries, most notably in China where some 1.4 million electric cars were sold in 2019, compared to almost zero in 2014.[5] More and more manufacturers are eager to look for alternative production materials, reduce their energy consumption, and reuse components to make new products.

Most of them may have noble intentions in doing this, but even if they don't, they no longer seem to have much of a choice. Regulation is getting stricter; big economic blocs such as the European Union present real plans with real targets.

The EU wants to be climate neutral and have a 'circular economy' by 2050:

> 'In 2050, we live well, within the planet's ecological limits. Our prosperity and healthy environment stem from an innovative, circular economy where nothing is wasted and where natural resources are managed sustainably, and biodiversity is protected, valued and restored in ways that enhance our society's resilience. Our low-carbon growth has long been decoupled from resource use, setting the pace for a safe and sustainable global society.'[6]

How exciting is that vision of the planet in 2050? And how would it translate to the world of cycling?

Imagine a bike that has been made from plant-based materials or recycled and reused parts, and that the material wearing from your tyres or brake pads is biodegradable. The lubrication washing down from your chain no longer pollutes the forest you are riding through, but provides valuable nutrients for the plants in it. Sensors tell your cycling computer about the state of the components and warn you when and how to maintain them. You no longer discard your old bike as if it were a piece of rubbish, but return it to the manufacturer so that parts and materials can be reused to make new bikes; the condition of the components has been monitored by the data sensors and collected in the bike's own bike passport. Or, alternatively, you could plant your old bike in your garden for it to become part of the circle of life again.

You would be living in a world with a circular economy. A world where you ride your bike in an environment without pollution. Through forests larger than today, inhabited by ever more varied species of plants and birds. A world where CO_2 emissions no longer contribute to climate change and we no longer dig up finite resources from the Earth, but use our 'waste' or renewable natural sources to make new products.

This vision of a Circular Cycling Industry is what drives us. To make the transition from our current linear economy of 'Take, Make and Waste' to a circular economy, marginal gains are not sufficient. We will have to change the way we design products, business models, and the interaction between manufacturers and the users of products.

This book is a practical guide to help the world of cycling make that transition. We hope to inspire people in the cycling industry, designers, business developers, marketeers, event & race organisers and governing bodies, as well as riders, to transform the way we think about bicycle design and the use of a bicycle.

The cycling industry has proven to be a very innovative industry, able to deliver better products every season by teams based all over the world. The supply chains are truly global: design is done all over the world, manufacturing mainly in Asia, customisation closer to the customer. The internet makes it possible for brands to interact directly with consumers, who now no longer rely solely on their local bike shops to buy bikes and parts.

Bike innovation has made a huge difference: road bikes have become more reliable through better component design, testing and manufacturing methods. Shifting and gear ratios have improved, ergonomics of handlebars and saddles have made riding a bike far more comfortable, making the road bike accessible to more and more people across the globe. In fact, the design of the road bike has reached such a high level of maturity that the number of radical breakthroughs has been very limited in recent years. It is no coincidence that the term 'marginal gains' has been introduced in the industry, since you now need a lot of tiny steps to make a difference. It is a fitting expression, not just for the way (pro) riders become better riders, but also for bike design these days.

However, the industry's focus on standing out by introducing marginal innovations in order to sell more products, and sell them faster, is no longer sustainable. This approach causes too much waste, as more and more products will cease to be used and eventually discarded before they have reached the end of their technical lifespan.

The transition to a circular economy offers an enormous opportunity for bicycle brands to make a big step in product design and in their interaction with customers. It requires the introduction of radical innovations, leading to a larger market share for the most innovative companies. Some readers might remember how the race bike has evolved since the steel bike Indurain rode in 1992. In the 1990s and early 2000s, cyclists all over the world eagerly awaited the innovations every brand would introduce in next year's models. Innovations that really made a difference in bikes' performance and gave consumers a real choice in what to buy.

The introduction of bicycles for a circular economy will be just like that: exciting, with a lot of new products and services to choose from. There will be successes and there will be failures. Bikes will be better – not just better to ride, but better for our planet and our societies. It is time for the cycling industry to adopt business models and product designs that no longer deplete natural resources, cause pollution and CO_2 emissions, and leave so many high-tech materials to end up in a landfill or an incinerator.

To get there, we need much more than marginal gains, we need a revolution.

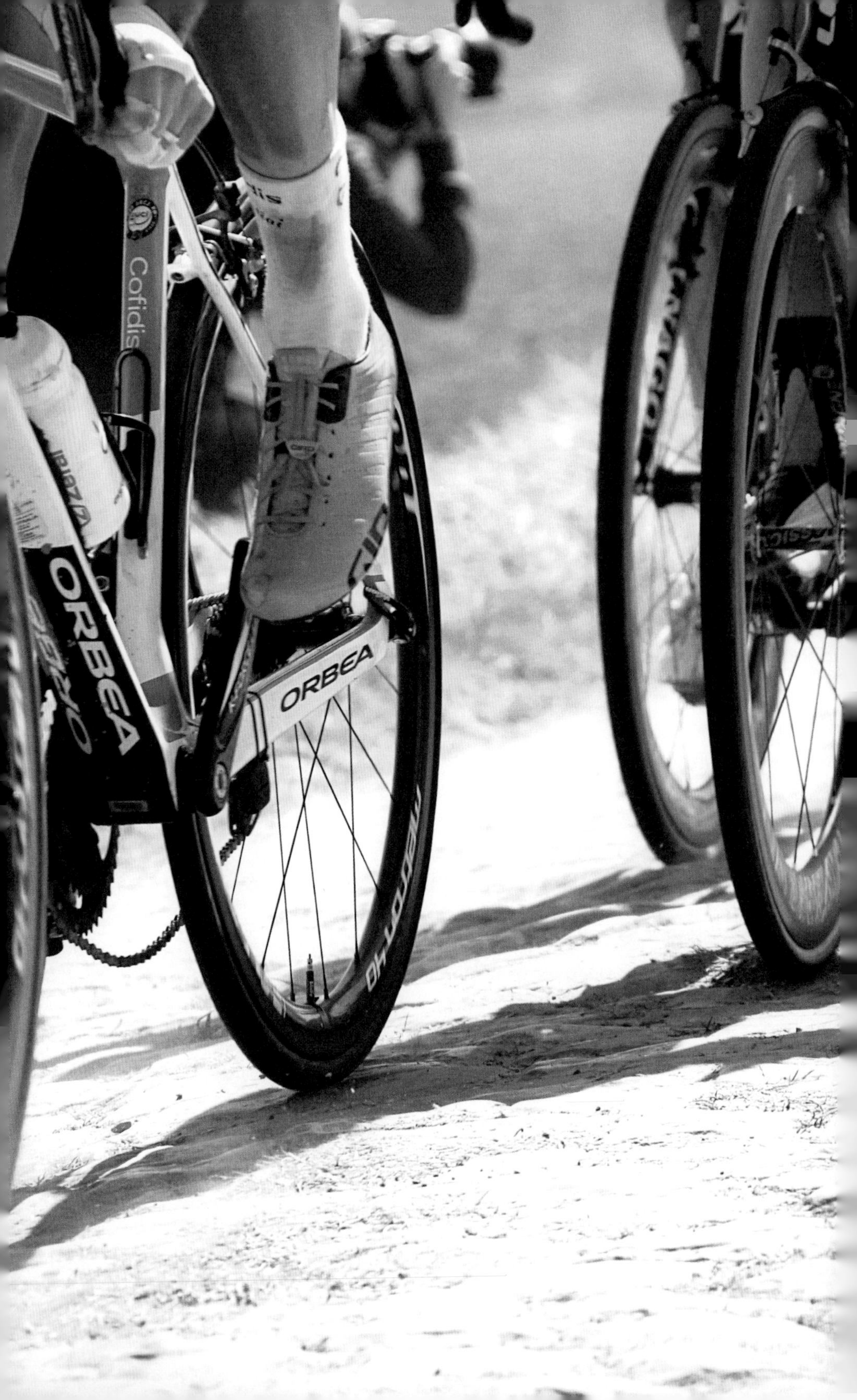
Cofidis
ORBEA
ORBEA
ORBEA

THE ROADBOOK TO A REVOLUTION

In your shed or attic there is a Box, a box filled with bike parts. They used to be on one of your bikes, or you bought them as spare parts but never used them. Maybe you got parts from a friend, because, you know, they might come in handy one day. The parts are too good to just throw into the bin, but deep down inside you know you will never use them again. Then you move to another house, you come across the Box and have to decide what to do with it.

With a bit of nostalgia, you go through the parts: 'Remember, that beautiful and super durable Chris King NoThreadSet™ that has survived three of my bikes? And – oh wow, back then I had several pin-ups to put on my Cinelly stem just like Mario Cippolini. And here is a 9-speed chain that I bought as a spare but that does not fit on my current bike anymore.' In the end, you decide to get rid of the parts, because after years in the Box they are no longer compatible, technically outdated or simply out of fashion.

Does this sound familiar? Think about it for a minute. Is this really happening? Are you spending lots of money on stuff that you do not fully use and then discard, without considering the impact of your actions on our planet? You are not alone. Nearly all the cyclists (and shops, and distributors) we spoke to over the last few years have a Box. It is a typical symptom of the linear economy, where parts with a considerable remaining technical lifetime are stored, and eventually thrown away because new parts have replaced them.

UPCYCLES

Our dream is to ride the 'bike of the future.' One that adds nutrients to nature while we ride; a bike built without finite resources and without waste that ends up in the environment.

In the 1990s, we both started mountain biking as teenagers. Our local single tracks were full of sticky mud that caused our bikes to wear very fast, but we had no money to have our bikes repaired or to buy new parts. We learnt to maintain and repair our bikes early on, simply because we could not afford to buy replacement stuff all the time. Being bike nerds, we enjoyed the era of rapid improvements in bike technology. We worked in bike shops alongside our studies, spending every euro we earned on our bikes to make them better, or to buy another one.

Even after finding full-time jobs, we kept riding our road, gravel and mountain bikes. We also continued our conversations about the way the cycling industry works, influenced by what we learnt every day – Matthijs as an IT manager inside the industry and Erik as an innovation manager outside the industry. In the late 2000s, sustainability became an important part of the strategy of many companies, initially aimed primarily at reducing negative impact. Erik got involved in some of the first projects in the Dutch construction industry that were based on ideas about a circular economy, where reducing environmental impact and making money go hand in hand.

In 2018, we decided that just talking about how the cycling industry was not picking up this trend was no longer good enough. We decided to act, and founded Circular Cycling, a commercial start-up aimed at transforming the waste in the Boxes into potential resources. It was an experiment to test a few tiny first steps on the road towards a circular economy: we built and sold UpCycles – new bikes made from used parts. At first, we used the parts that came from our own Boxes, then from numerous other Boxes.

Each and every UpCycle was unique, as we had the opportunity (and needed) to mix and match parts and framesets in a way we thought fitted best together. We mounted new handlebars (safety first!), bar tape (someone else's sweat doesn't really feel new does it?), cables and comfy 25mm tyres on each bike. Drivetrains were checked for wear and replaced when too little life was left in them. Once finished, the 'refurbished' road bikes were offered to consumers for prices some 30%-50% lower than you would pay for a new model with the same ride quality.

We made many mistakes and were in no way a perfect example of a truly circular or sustainable company. We eventually stopped building and selling the UpCycles, because we ran into the limitations of using products that are designed for a linear economy in a circular business model. However, we did learn many valuable lessons about the complexity of the cycling industry. Our experiments showed us how a transformation of the current linear system into a circular cycling industry could look like.

To realise our dream of a circular cycling economy, we found that the entire industry needs to change. We share our insights in this book because an experiment in a small shop in the Dutch city of Utrecht is not a revolution. For a revolution to happen, small initiatives like ours that are already happening all around the world, need to connect with each other and with new initiatives to create enough momentum to start a revolution on a global scale.

A CIRCULAR BUSINESS MODEL INCLUDES THE ENTIRE SYSTEM

We talked to many people as part of our research for this book, including publishers and marketeers. Almost everyone asked us who would be the target audience of this book. 'Are you targeting readers looking for a management book about business or engineering? Or a book about cycling? Is it about sustainability?' The answer is yes – to all those questions. Most people told us that we needed to focus, 'target one group' they kept telling us. But if you want to change a complex system like the cycling industry, we believe that you need to address it in its entirety.

Bike and parts designers, manufacturers and their shareholders, marketeers, rule makers (the UCI and National Federations), event & race organisers – they are all part of the current linear industry model, making profits in a way that has been perfected over decades. And cyclists (let's not forget the consumers!), who ultimately make the decision what to buy, when to buy and how to take care of their gear when they go out on the road. They might not feel it that way, but they too have an enormous impact on the way the industry works.

The good thing is that all these different groups have two things in common:

1. They are all human beings who can make a difference if they want to: they can all make a conscious decision on what to design, how to invest and what to buy. Or on how to set rules for the future technical (or maybe ecological?) requirements of a bicycle competing at, say, the 2028 Los Angeles Olympics.
2. They all live on the same planet with its limited resources and changing climate.

In this book, we have tried to address all the stakeholders in the cycling industry, and therefore to avoid too much cycling and circularity jargon. Some bits might be more interesting to you than others, depending on your role in the system (for example on whether you are a cyclist with an interest in sustainability or a designer of framesets).

HOW TO READ THIS BOOK

The information we present in this book is a mix of the experience we gained getting our hands dirty building our UpCycles, but also designing circular products in the construction industry, training companies on circular business models, as well as managing the IT of a bike distributor, and developing online databases for digital product information. We have included relevant scientific research done by others on topics such as circular economy and climate change.

This book is like a one-week stage race: the preceding prologue will be followed by five stages, all with a different character. A book short enough to race through, but long enough to see the landscape change along the way.

STAGE 1   THE COMPLEX WORLD OF CYCLING AND SUSTAINABILITY
STAGE 2   THE LINEAR CYCLING ECONOMY
STAGE 3   CREATING VALUE IN A CIRCULAR ECONOMY
STAGE 4   ACTION PLAN FOR A CIRCULAR CYCLING INDUSTRY
STAGE 5   GETTING THERE
FINISH   STAND UP FOR A REVOLUTION

We will present you with a mix of facts about the challenges resulting from a linear economic system, the way circular business models might work, and examples from inside and outside the world of cycling. After reading this book, you should have a better idea about a circular economy, what this could look like in the cycling industry and what you can do from your position in the world of cycling to contribute to the revolution.

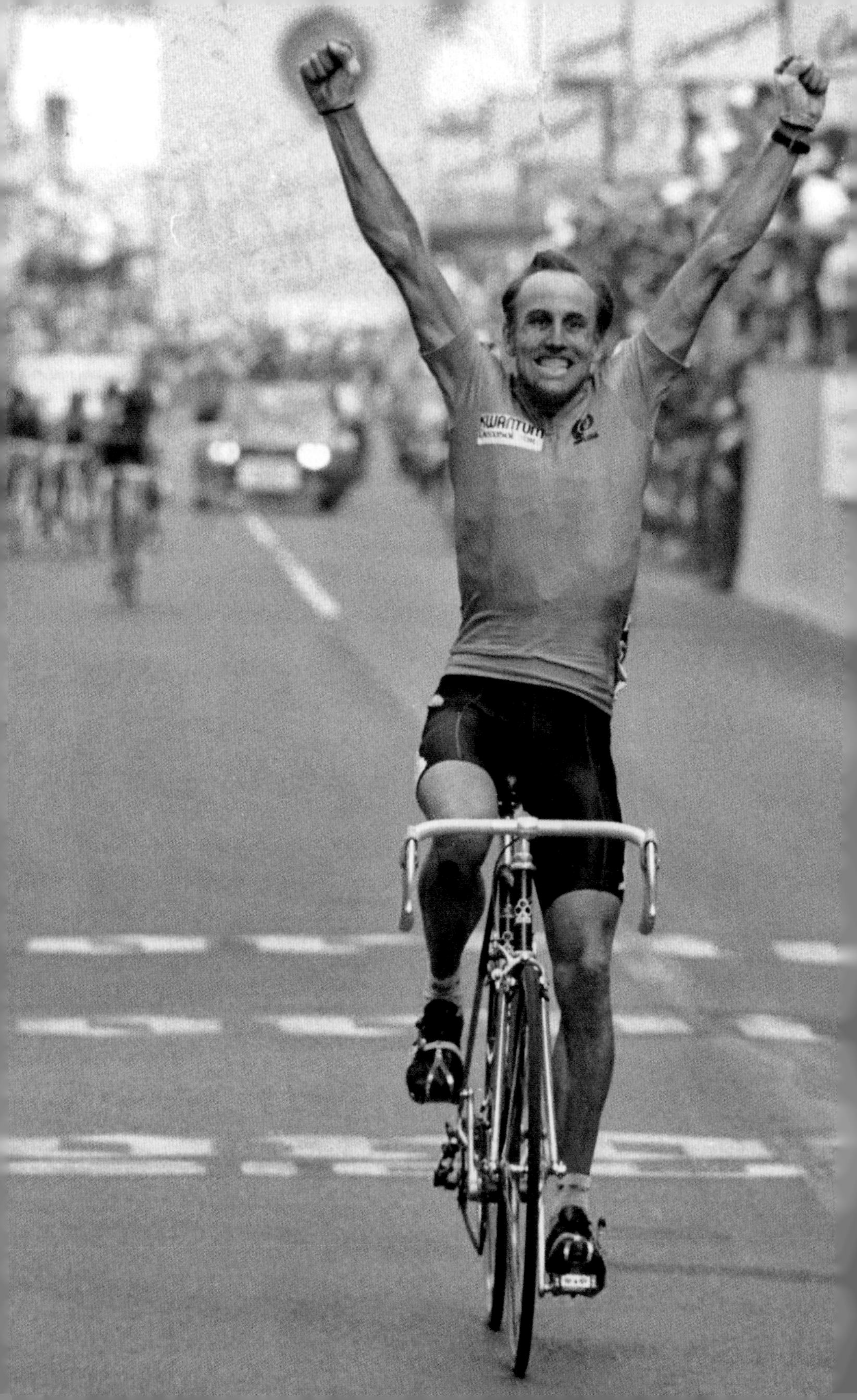
KWANTUM

STAGE 1
THE COMPLEX WORLD OF CYCLING AND SUSTAINABILITY

On 1 September 1985, an estimated 300,000 spectators lined a 14.75km course in and around Giavera del Montello, just north of Venice in Italy. They came to see bike riders pass 18 times in an effort to win the UCI World Championships. It was not until the final climb on the final lap that the decisive breakaway was formed, after several riders including Stephen Roche and former world champion Greg Lemond attacked. At the top, only 14 riders were in a position to win the race. The Dutch had Johan van der Velde, Gérard Veldschoten and Joop Zoetemelk in the breakaway group, the Italians had Moreno Argentin and Claudio Conti, and the other nine riders had no help from teammates.

Zoetemelk, 38 by then, was renowned for his many second-place finishes in the Tour the France behind Bernard Hinault and Eddy Merckx. No one really believed he was the man to beat that day, including himself. He worked hard for the team to chase riders that tried to escape. Van der Velde was a decent sprinter, so it made sense to sacrifice Zoetemelk to make sure Van der Velde could go for the sprint. With about three kilometres to go to the finish, Zoetemelk came up from the back of the group to take the front position again to prevent others from breaking away. He passed the group so fast that by the time he took the front, he immediately had a small gap. Not a single rider made a real effort to reel him back in; everyone looked at someone else to do the dirty work. The Italian couple was forced to put Conti at the front of the group, but a single rider chasing down a rider as strong as Zoetemelk on his way to winning his first World Championship did not stand a chance. When Conti looked over his shoulder to ask for help from the other riders, not one rider made a real contribution. Zoetemelk won.

1.1 BIKE DESIGN, RULES & REGULATIONS

The Colnago bike Zoetemelk rode in 1985 represented the state of the art of technical development back then. All the mastery, tricks and wisdom that Ernesto Colnago was renowned for had come together in that bike. Yet at the same time, it did not differ significantly from the road bikes of the 1950s. The frame and forks were still made of steel tubes and lugs with about the same diameter. And just like in the 1950s, road bikes were equipped with down-tube-operated derailleurs and rim brakes. Wire spoked wheels, pneumatic tyres and curvy, dropped handlebars had all been around since long before World War II. Of course, bikes were more reliable, weighed less, their stiffness had increased, and the number of gears had grown by 1985, but all in all, the differences were not that big.

It was not until the 1990s that things started to change drastically. French bike manufacturer Look introduced 'clipless' pedals where the shoes click into the pedals, similarly to how ski binds worked around the time Zoetemelk won the World Championships. In 1990, Shimano introduced the Dura Ace STI brake levers with integrated shifters; Campagnolo followed the next year with a similar product, which meant that riders no longer had to reach for the downtube to change gears. Overall weight continued to go down and even more gears were added, but frames were still mainly made from steel, the geometry of the bike was more or less fixed, and wheels looked almost the same as in the 1950s.

In 1985, the way most middle to high-end road bikes were sold was quite different from how they are sold today. As a consumer, you went to your preferred bike shop, where you would select a (steel) frameset from a catalogue or from a rack hanging off the ceiling. Components were handpicked and buyers sometimes selected used components that were in good condition to save costs. Once the frame and parts were delivered, the shop would build the bike for you – it was a one-off, custom bike. Both our dads spent many hours at their local trusted bike shop debating options for their new bikes with the shop owner.

The reason for this elaborate buying process was quite simple: until the 1990s, it was not possible to mass produce high-quality road bikes for a price that offered a real advantage to the old artisanal model. Coinciding with the introduction of aluminium framesets, this changed completely as new manufacturing methods and management strategies were introduced. Mass production took an enormous flight, and mass-produced road bikes both became cheaper and increased in quality. Overall, the quality of components made in Asia improved and the standard 'rotten apple' components found on complete bikes (think of headsets, seatposts and hubs) were weeded out. Wheel quality went up, with wheels staying straight and the spokes not needing re-tensioning time after time. In-house brand components, such as seatposts and integrated handlebars, evolved from second-tier products with a logo, to parts designed to work specifically with a frameset. Finally, these mass-produced bikes were backed by fantastic marketing strategies via ever more channels – in fancy colour brochures at first, then online.

The shift from small-scale production of frames and components that were assembled to complete road bikes at the bike shop, to mass-produced high-quality complete bicycles took place over a period of about ten years. The impact on the industry was enormous. In local bike shops, focus had to shift from making money by building and repairing custom bikes, to selling mass-produced bikes that were delivered to the shop in big boxes, almost completely assembled. The business model shifted from profits on labour to margins on sales. Small-scale frame builders struggled to offer complete bikes at a competitive price, because they lacked the scale of the big brands.

Some managed to make the transition, others did not as American brands started selling their mass-produced bikes on the European market. Also, new European, mostly German, brands stood up to offer new bikes with a never-before-seen quality/cost ratio.

The new brands outsourced more and more work to factories in Taiwan and later China, to such an extent that practically all components are now produced in Asia. The supply chains of manufacturers involved in making a road bike have become longer than ever before. Big brands now have over 200 suppliers, each with its own sub-contractors, who may have their respective sub-contractors, and so on. The cycling industry is now truly global.

Quality control played a major role in this shift towards mass production. During the 1980s and 1990s, many frames and bike parts were taken to market without proper testing, and frames and parts too often failed while riding. Designers and suppliers were not familiar with the properties of the new materials and production processes that were used to design and manufacture all sorts of innovative parts. It was up to the end users to properly test innovations, while designers were already working on their deadlines to release next year's models. Many brands had serious warranty and reputation issues, leading to reduced sales and financial losses. Since the introduction of the European Standard for 'Safety requirements and test methods' (EN 14781:2003) in the early 2000s, the quality of products leaving the factory has improved dramatically.[7] The standard specifies the way parts and frames need to be designed and tested from a safety perspective, including stiffness, impact and fatigue tests. The result is that, outside crash situations, frames and parts nowadays hardly fail suddenly, due to improved designs and rigorous testing regimes.

The required investments in order to set up the quality control systems and a global supply chain, required external shareholders in what used to be fairly small companies. To be able to generate the required returns on investment, brands adopted a strategy based on growth. The market did indeed grow over the past three decades, and at the same time became fiercely competitive. Just like in other industries that switched to mass production, new strategies were required to keep their promises to shareholders.

The first one, adopted by virtually all big brands in the industry, was the introduction of model years. While frames remain unchanged for three to four years, brands introduce a completely new collection every year with, for example, updated groupsets, new colours and some minor component changes. Sometimes the price is adjusted to match or slightly undercut the competition, or to make up for changing currency exchange rates. Bikes are produced in limited series, which forces bike shops to pre-order bikes for the whole year to have a guaranteed stock. This often leads to excess stock, which has to be put on sale when the next model year is pushed by brands' marketing machines.

The second strategy is to look for ways to stand out from the crowd. Now that quality no longer is a real issue and the global supply chains are in place, bike designers have run out of real opportunities to set themselves apart from the competition. An additional major reason for this is that the basic form of the road bike is dictated by the international 'government of cycling', the UCI. In its 'Approval Protocol for frames and forks',[8] the UCI dictates the boundaries of bike design from a geometry point of view, in order to create a safe level playing field for all riders entering a cycling race. Because manufacturers prefer not to make separate designs for races and for general customers, the Protocol has an effect on all road bikes. Bike companies design their top model framesets catering for the needs of elite riders. Marketing focusses on the top models their sponsored pro riders use, as this is the most exciting story to tell, and consumers love to mirror themselves to the pro riders. Lower range bike models are also largely based on these top models, maybe with a bit more relaxed geometry here and there.

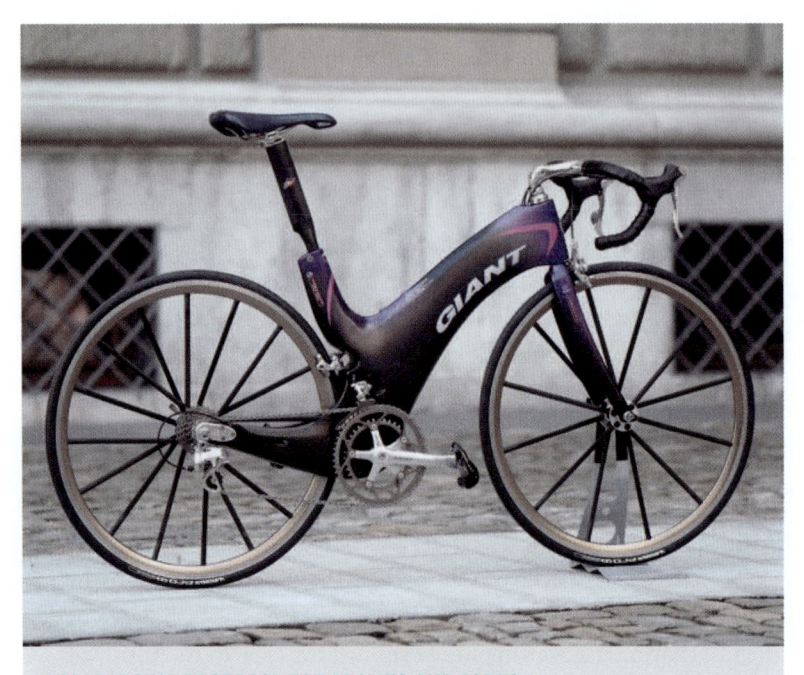

UCI APPROVAL PROTOCOL FOR FRAMES AND FORKS

In the late 1990s, bike brands such as Giant and Pinarello introduced some of the most fascinating frames ever sold. The futuristic Giant MCR (pictured) and Pinarello carbon monocoque time trial machines offered the riders a significant aerodynamic advantage over traditional bike designs.

Within a few years after these bikes became available and were used in competition, the UCI updated its rules to ban these exotic designs, based on the idea that:

'Bicycles shall comply with the spirit and principle of cycling as a sport. The spirit presupposes that cyclists will compete in competitions on an equal footing. The principle asserts the primacy of man over machine.'

The boundaries of bike design are laid out in a series of articles in the UCI Approval Protocol, for example article 1.3.020, which limits the size and aerodynamic profile of frames and forks:

Configuration
For road competitions other than time trials and for cyclo-cross competitions, the frame of the bicycle shall be of a traditional pattern, i.e. built around a main triangle. It shall be constructed of straight or tapered tubular elements (which may be round, oval, flattened, teardrop shaped or otherwise in cross-section) such that the form of each element encloses a straight line. [...] The maximum height of the elements shall be 8 cm and the minimum thickness 2.5 cm. The minimum thickness shall be reduced to 1 cm for the chain stays (6) and the seat stays (5). The minimum thickness of the elements of the front fork shall be 1 cm; these may be straight or curved (7). [...] The top tube may slope, provided that this element fits within a horizontal template defined by a maximum height of 16 cm and a minimum thickness of 2.5 cm.'

Since 2011, manufacturers can apply for the approval of a new frameset or fork every year before 30 June, and if approval is granted by the UCI technical committee, frame and/or forks can be used in competition from 1 January of the next year.

UCI

APPROVAL PROTO
FOR FRAMES AND FO

▷ **1.3.011 a) Measurements** (see diagram «Measurements»)

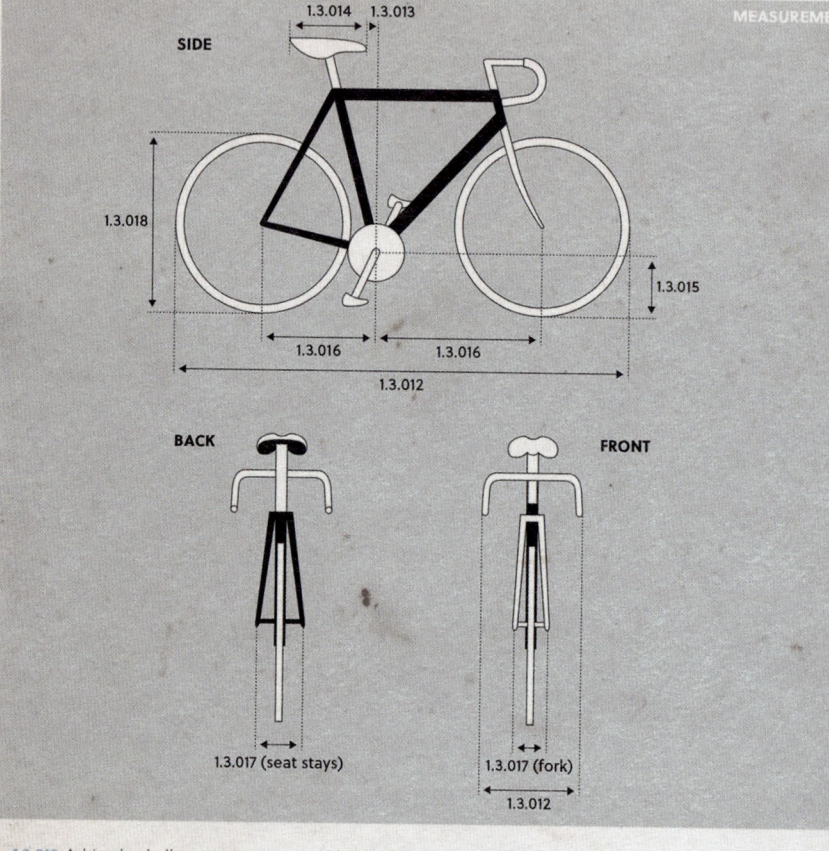

MEASUREMENT

SIDE
1.3.014 1.3.013
1.3.018
1.3.015
1.3.016 1.3.016
1.3.012

BACK
1.3.017 (seat stays)

FRONT
1.3.017 (fork)
1.3.012

▷ **1.3.012** A bicycle shall not measure more that 185 cm in length and 50 cm in width overall. A tandem shall not measure more than 270 cm in length and 50 cm in width overall.

▷ **1.3.015** The distance between the bottom bracket spindle and the ground shall be between 24 cm minimum and maximum 30 cm.

▷ **1.3.016** The distance between the vertical passing through the bottom bracket spindle and the front wheel spindle shall be between 54 cm minimum and 65 cm maximum (1).

The distance between the vertical passing thro the bottom bracket spindle and the rear wheel sp dle shall be between 35 cm minimum and maxim 50 cm.

▷ **1.3.017** The distance between the internal extrem of the front forks shall not exceed 10.5 cm; the dista between the internal extremities of the rear triar shall not exceed 13.5 cm.

UCI TECHNICAL REGULATIONS FOR FRAME AND FORKS

APPROVAL PROTOCOL
FOR FRAMES AND FORKS

SHAPE 1

CROSS SECTIONS

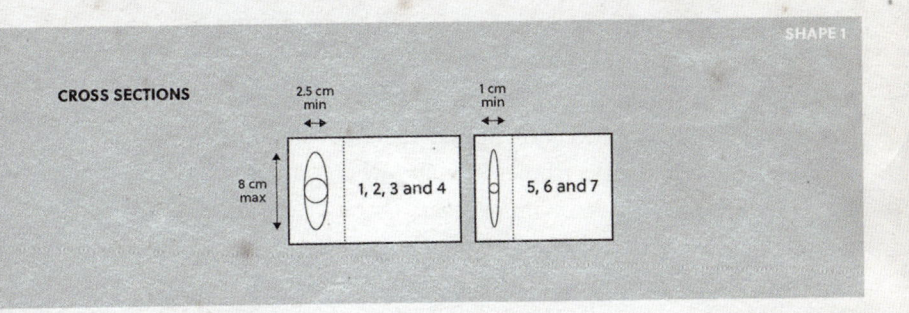

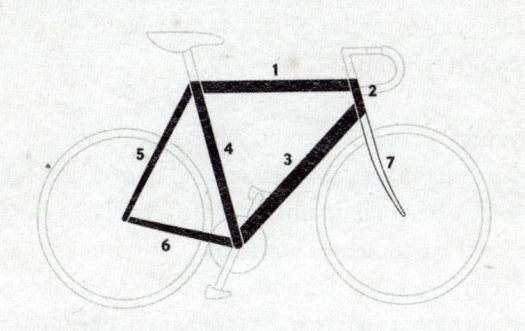

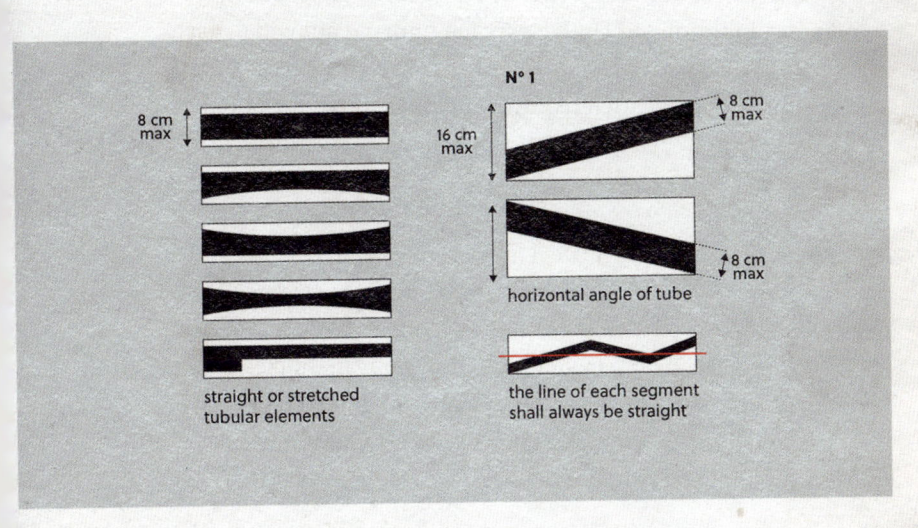

2.5 cm min — 8 cm max — 1, 2, 3 and 4

1 cm min — 5, 6 and 7

N° 1

8 cm max

16 cm max — 8 cm max — 8 cm max

horizontal angle of tube

the line of each segment shall always be straight

straight or stretched tubular elements

To differentiate their product, designers search for tiny improvements wherever they can be found, leading to so-called marginal gains. Often these gains are realised by designing a part that does not fit existing standards, such as bottom bracket sizes and seatpost shapes specific to a certain frameset. The result is that all sorts of interface solutions are needed to make components work together, similar to the electronics industry where you need different plugs and different cables for almost every single device and country. The only real standards for bike components that seem to be excluded from the search for marginal gains, are the dimensions of the tyre/rim interfaces, the thread in cranks to accommodate for pedals, the dimension of saddle bridges, and the diameter of a handlebar to accommodate for shifters clamps.

Both the mass production of bikes and the increase in different components that are no longer compatible, lead to over-production of many parts, which often end up unsold. Matching large numbers of bikes with the right customers is hard, matching stocks of all sorts of unused interfaces and parts with rare fittings to bikes that actually need them is even harder. As we experienced first-hand when building our UpCycles, even figuring out which parts were compatible with a frameset and with each other was a time-consuming process, which required not only sufficient experience to do the research, but also a vast stock of all sorts of different parts to make everything work together.

Selling products that are pushed into the market to reach sales targets, requires clever marketing to make marginal performance gains look like the best innovation the cycling world has ever seen. And we, normal consumers, fall for that. We buy new bikes, new wheels, new clothing, new computers, anything new because marketing tells us that the latest products are the best, the fastest, the most compliant and simply essential if we want to be better riders.

We only too often forget that cycling is about having fun, about being outdoors and that our legs have far more impact on our performance than the latest aerodynamic improvement of a bike part. In addition, most of the time we ignore the impact of our consumption on the environment.

1.2 THE TRAGEDY OF THE COMMONS

Back to Zoetemelk's magic victory in 1985. Zoetemelk just wanted to ride tempo at the front of the group, got a small gap and that was that, somehow. What happened? The chasers were caught in what is called a tragedy of the commons: if every rider in the group had taken a few short pulls at the front of the group, they would surely have caught up with Zoetemelk and one of them could have been world champion. Instead, no one wanted to invest any energy, afraid of losing to someone else in the group.

The theory of the tragedy of the commons was described in a paper by Garrett Hardin in *Science* in 1968.[9] The common is a shared-resource system such as an open field in a village where, according to local English and Irish traditions, local herdsmen were allowed to let their cattle graze temporarily, making sure that the shared resource would not be depleted. Each herdsman had a responsibility to ensure there would always be enough grass to feed the others' sheep as well.

Short-term self-interest dictates that each herdsman maximises his own gain by grazing as many of his sheep as possible on the commons. Perhaps even a few more than his agreed right, even though he knows it is not a good thing for the community as a whole. This is called 'freeriding'. If all herdsmen act in that way, overgrazing soon becomes a reality and the abundant pastures of yesterday are turned into a muddy field with hungry sheep. The common good suffers from the freerider behaviour of a single individual. In Hardin's words: 'The individual benefits as an individual from his ability to deny the truth even though society as a whole, of which he is part, suffers.'

In the chasing group of the 1985 World Championships, all riders were literally hoping for a 'free ride': energy they would have to spend on raising and maintaining the speed of the group to bring back Zoetemelk was spared to maximise individual chances of winning if it would come to a sprint finish. The tragedy here is that the speed of the chasing group was not properly managed and that a limited resource was wasted: the group lost the opportunity to win the race.

The world of road cycling is full of situations like this; stories about broken informal rules and chasing groups derailed by the blatant self-interest of one or more individual riders. Of course, it is part of what makes cycling such an exciting sport to watch. The 'real world' is no different, although unregulated resources may often be more distant there, and therefore more abstract. In our linear economic system, there are many 'commons' suffering from freeriding behaviour, with huge and very real consequences. Examples include the oceans (overfishing leading to reduced fish populations), the atmosphere (emissions causing air pollution and climate change), and finite material resources (leading to scarcity for future generations).

Managing common resources has been part of human society for as long as we have been around. The balance of accepting lower personal profits in return for a reduced risk of disaster for all, existed long before there were written laws, governments and nation states. Basically, there are two ways to manage a common resource: with and without a government.

The non-governmental, bottom-up option is to organise the resources with all parties involved through a dialogue where each and every party understands all others' point of view and a suitable outcome is reached together. This works well in situations with a clear common resource, a strong social network capable of sanctioning a freerider, and a direct benefit of the resource for the local community – for example the water supply of a small and remote village. When communities get too large to remain a strong social network, or the effects of the degradation of the resource are unevenly spread, a top-down government type of organisation is required. It can either implement regulation or 'internalise negative externalities', basically taxing negative effects such as pollution.

1.3 GOVERNING THE CYCLING INDUSTRY

In the world of competitive cycling, one finds both solutions. Because top-down regulation is impossible for the safe navigation of every imaginable situation in a race at 50km/h on sometimes dangerous roads, riders in a peloton ensure a minimum level of safety amongst themselves. Because all the riders directly benefit from some sort of ruleset, the social structure of a peloton is small enough to discuss safety internally, and riders who break the rules theoretically can be punished by the rest of the peloton ('whatever happens, together we will stop this guy from winning'), the common resource (in this case: safety) is managed quite well.

However, the peloton needs a 'cycling government' as well. Rules are required to make the sport fair, safe and financially sustainable. Many rules, big and small, have been developed over the last decades, ranging from qualification systems for races to the use of rainbow stripes on clothing and sock height. The UCI does this on a global level, while National Federations and race organisers implement rules on a local level, in line with specific circumstances. In general, this system works very well, although recent history has provided us with a complex tragedy of the commons in cycling that undermined the entire sport: the widespread use of doping in the 1990s and 2000s. It was probably the largest and most painful crisis in the history of the cycling sport – but it also shows the ability of the collective cycling community to clean up its own mess.

Just like climate change and resource scarcity today, the extensive use of doping was an inconvenient truth. Everyone in the cycling community seemed to be aware of the problem, from individual riders to teams, doctors, race organisers and cycling federations. Even journalists, sponsors and spectators adoring their heroes knew what was happening, but they were in denial and kept supporting their idols. Initially, no one dared to speak up because the individual gain (of using doping to increase chances of winning races), was larger than the individual discomfort (of the damage done to the sport as a whole).

Especially after the scandalous Tour de France of 1998, nicknamed 'Tour de Dopage', when French police raided several teams and found large quantities of doping, more effort was put into fighting doping. In 1999, the World Anti-Doping Agency (WADA) was set up, an independent global agency with funding and power from outside of cycling. After the 2006 Operación Puerto scandal, when blood bags from many stars of the peloton were found in a fridge near Madrid, seven pro teams started the Mouvement Pour un Cyclisme Crédible (MPCC), and in 2008 the Biological Passport was introduced to improve the monitoring of riders.

Eventually, journalists, sponsors and the public, as well as more and more riders themselves, wanted the use of doping to stop and the sport did get cleaner. Better rules, stricter policing, tougher sanctions, a new generation of riders and many other initiatives, broke the reasoning that all riders needed doping to be able to compete. In the past, it was the guy who did not use doping who was the exception. Now it seems to be the rider who does use it.

As we will explain later, we will need a similar collective effort to escape from the linear cycling economy. Together we are stuck in a tragedy of the commons where every stakeholder in the cycling world will have to invest something to tackle the issues arising from the linear cycling industry. The number of players involved in the entire cycling world, from regulators and people working on the production line of framesets in Asia all the way to consumers, is large and spread all over the world. The stakeholders also have a very diverse set of (short-term) interests. The UCI has the power to set rules, but the other stakeholders cannot be forced to implement these outside race situations. Manufacturers will have to find ways to make money to cover the short-term investments they have to make. Riders will have to start realising that a new bike every couple of years is not a sustainable consumption pattern fitting in a circular economy.

In this book, we will focus on road bikes when describing both the linear and circular economic models. With a little creativity, the same philosophy also works for other types of bikes such as e-bikes, as well as for cycling equipment and apparel. Event organisers and teams have to make designs and select materials for start/finish locations and marketing materials, and have to choose means of transport and catering – for them too, a circular model will be an opportunity to reduce their environmental impact. We all can and will have to contribute – waiting for someone else to do all the work will not bring us victory.

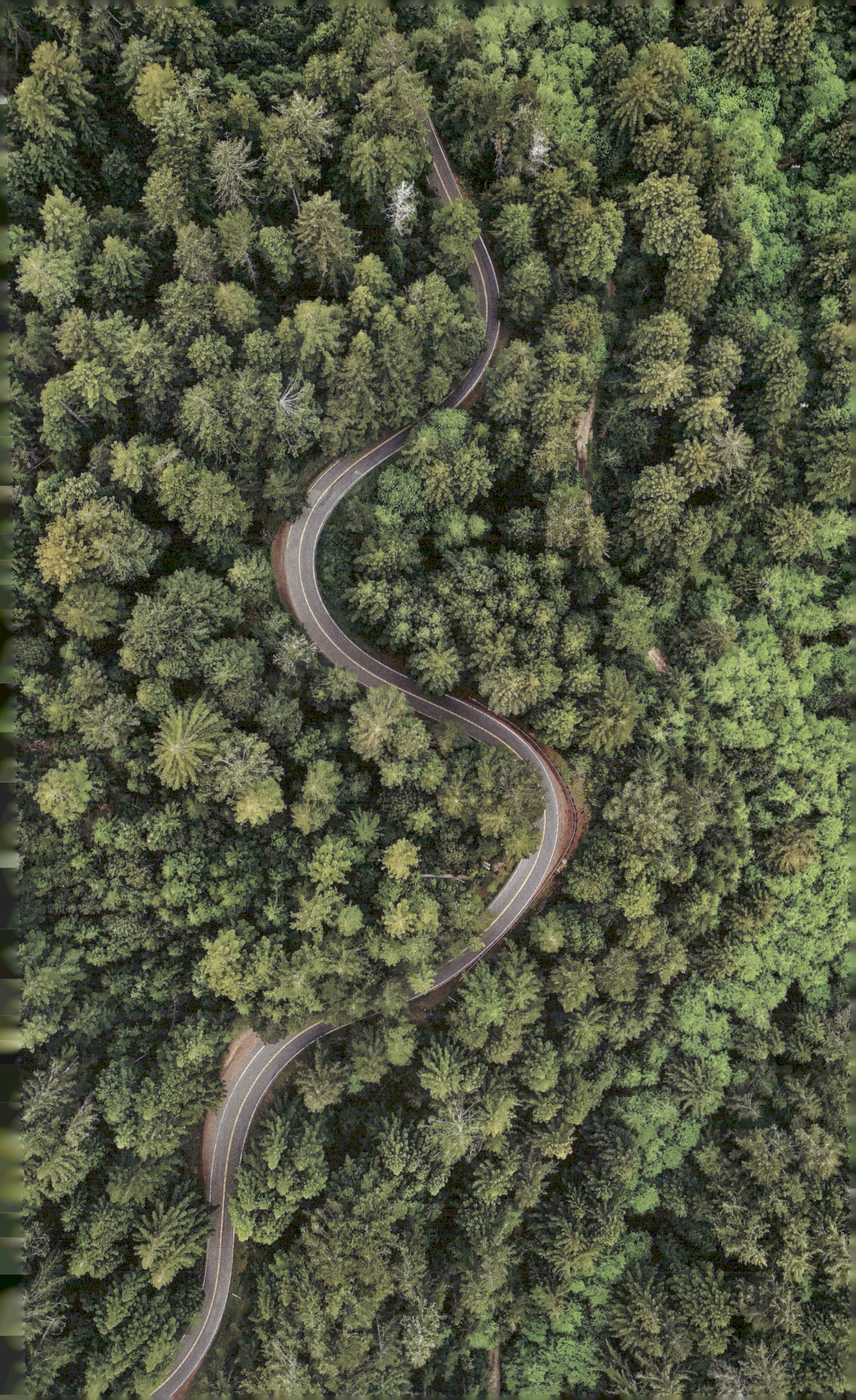

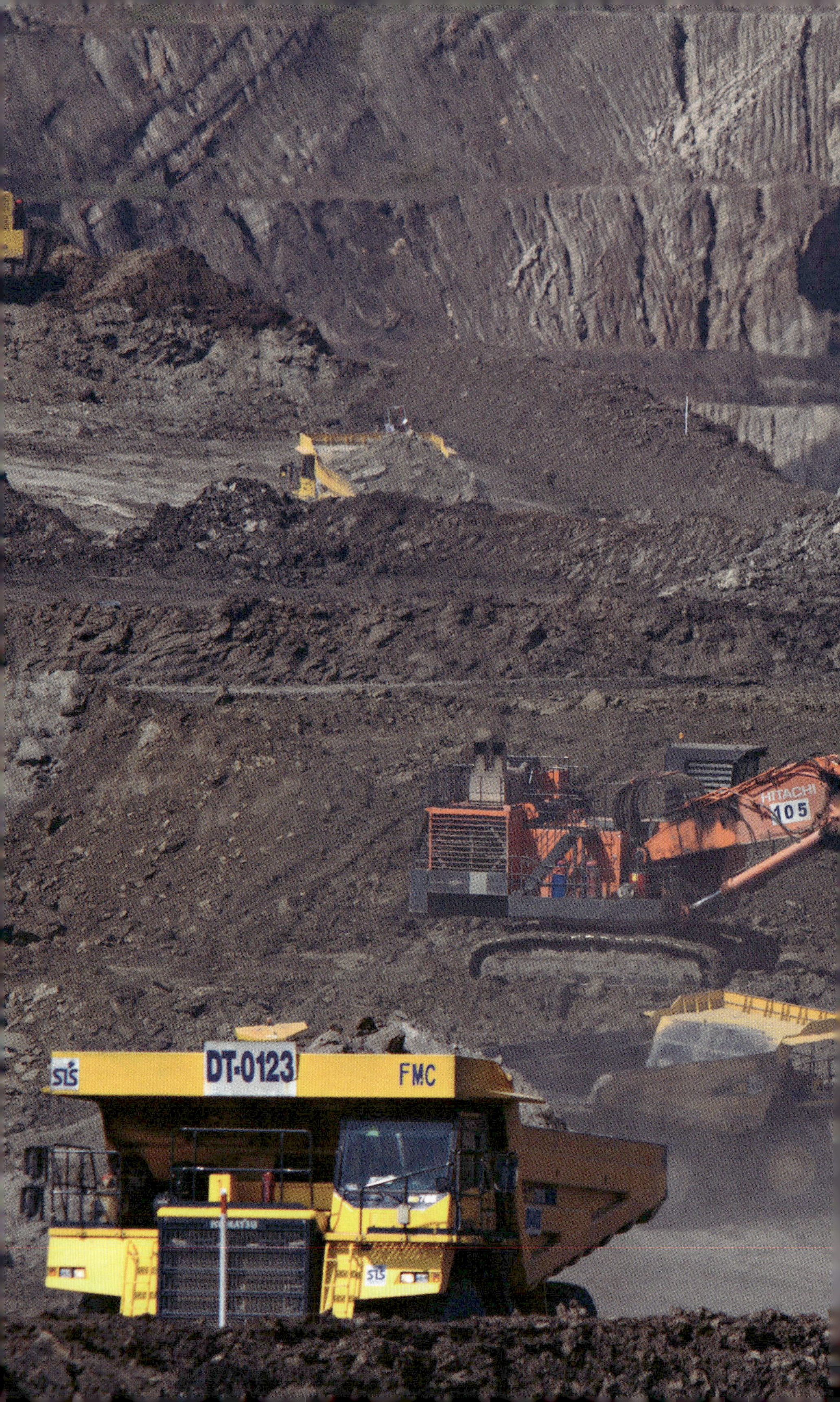
DT-0123
FMC
HITACHI
105

STAGE 2
THE LINEAR CYCLING ECONOMY

Sarah is a keen amateur rider. She rides some 3,000km a year, including a couple of Gran Fondos that she approaches as her yearly goals. After a number of seasons on a second-hand bike, Sarah found that cycling really was her thing and she decided to buy a fancy carbon road bike a few years ago. Her mates told her she needed to buy something lightweight for her Gran Fondo ambitions; a bike with a carbon frame and a high-end groupset. None of her mates were riding disc brakes and they all advised her not to go for this option, as it only added weight and was not a proven technology. Eventually, she bought a top-end model with a previous generation groupset that had been on show in the shop for a few years, with a 25% discount.

When she left the shop, Sarah was happy as well as anxious, never having spent this much money on a hobby. Yes, it is a super cool bike, but did she buy the right one? Would everything on the bike fit nicely, or would she be replacing the saddle soon for another more comfortable one? Would a bike without disc brakes still be cool in a few years' time, as she was reading in the magazines that these were going to be the new standard?

Sarah has now used the bike for a few years. Yes, she did replace the saddle for a different model after a couple of months, but other than that, the bike never had any issues, apart from the regular handlebar tapes, tyre and chain replacements. Now the new saddle all of a sudden started to sag, making it very uncomfortable. A close inspection showed a crack in the bottom of the shell. Speaking to her mates, she found out this was a common issue with saddles as well as other parts; she realised that a little extra material could have made the saddle last forever. Repairing it seems impossible, so she is forced to replace the saddle with a brand new one.

A couple of weeks later, Sarah crashes on a wet spot in a sharp turn on her local loop. It is not a serious crash; Sarah is not injured, and the bike has suffered no real harm. However, her handlebar hit the ground and her shifter is broken, the brake lever snapped off and the shifting mechanism somehow locked up. Sarah goes to her local bike shop that weekend to get the shifter repaired. It is busy at the bike shop and Sarah has to hang around for a while before someone is able to listen to her story. 'We're busy at the moment, but if you leave your bike here, we will check it out as soon as possible and give you a call.' Unhappy – the weather is exceptionally good this weekend and she had a couple of rides planned – Sarah leaves her bike at the shop.

Almost a week later, the shop calls Sarah and tells her that the shifter cannot be repaired and thus needs to be replaced, but that they cannot find a shifter of the same, older generation. They tell her that they need to order an entire new set from a new generation of shifters, and that this new generation is incompatible with her front and rear derailleur. This means that she needs to replace these too, and it will be another couple of weeks before the parts arrive and the shop has time to replace the old ones. The cost, including labour, is about 30% of the original price of the bike. The bike shop asks Sarah whether it might be an idea to buy a new bike with the latest technology instead? Even unhappier than before, Sarah decides to let the bike shop do the repairs.

Just after she gets her repaired bike back, she goes out on a training ride with a group of friends in preparation for her main event of the season. In a few weeks' time, they will head to the Alps together to ride some of the famous cols of the Tour de France. The six-day trip has different start and finish locations, taking them from Albertville to Nice.

Most of her friends seem to have bought new bikes this winter – shiny bikes with aero shapes and disc brakes are parked around the local coffee bar where they usually meet. Sipping her cup of coffee, Sarah listens to the stories being told around her. Disc brakes are a must-have it seems, the guys cannot stop talking about them. They all bought these bikes specifically for the trip this summer. 'The old types of brakes cannot get you down these mountains safely – if you go to the mountains, you need to have disc brakes,' is what she hears everywhere. Sarah listens and regrets her decision not to buy a new bike, instead of doing the repairs; her bike is clearly not as good as the new ones bought by the others.

2.1 THE LINEAR ECONOMY

Sarah's story is not a true story, but it easily could have been. What she experiences are issues typically related to the linear economic system that comes with the expectation of continuous growth of sales and revenues to generate as much 'shareholder value' as possible. Over the last decades, the cycling industry has developed a very cost-efficient system for the design, production, distribution and sales of products. It is a globally integrated network with some of the bigger brands having design offices in three time zones, and an array of manufacturers and suppliers in various (mainly Asian) countries. Labour costs, manufacturing expertise and capacity, environmental rules, import duties and other supply chain risks such as trade wars and pandemics, constantly require tweaks to the supply chain in order to keep costs down in a very competitive market. Marketing adds further value to the brand by selling its unique properties. The goal is to sell more bicycles and sell them faster at the highest possible margin.

We customised the Value Hill model[10] for the cycling industry to show how the value of materials changes over the life of a product. As can be seen in Figure 1, a linear supply chain can be represented by a three-stage product lifecycle, where the value of the materials used for the product varies over the lifecycle. In an uphill struggle during the 'pre-use' phase, the supply chain adds value to raw materials.

ADD VALUE

DESTROY VALUE

User

Retail

Assembly

Manufacturing

Extraction

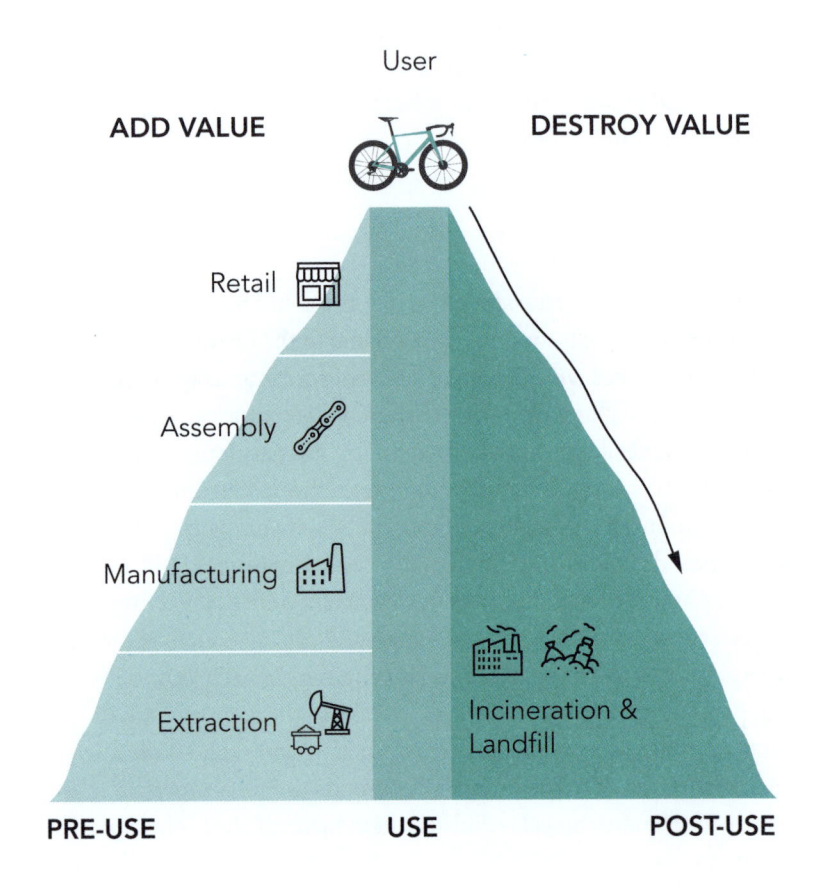

Incineration & Landfill

PRE-USE　　**USE**　　**POST-USE**

Figure 1: Linear economy Value Hill.

First through the extraction of materials from the earth, then the manufacturing of parts, the assembly of parts into complete bicycles, and finally the distribution and marketing of the products, before a user buys a product in a shop.

During the 'use' phase, the value – the ability to enjoy the bike for what it is meant to do, which represents the money the user is prepared to spend on it in the shop – is constant, assuming that the user maintains and repairs the bike at a reasonable level. At one point, often after a few years, the bike is no longer used and enters the final phase of the lifecycle. In the 'post-use' phase, the value of the bike goes downhill faster than the best pro riders on their way to a solo victory.

Our current economic system is not entirely linear. There is an economic reason and societal pressure to recycle at least some of the materials that we discard. But once the materials do reach a landfill or an incinerator, the value is gone.

2.2 LINEAR DESIGN INCENTIVES

One of the consequences of the linear economic system is that it creates incentives for brands to design products with a limited life span, to make sure that products 'fail' in a broad sense and consumers will buy new products to replace the broken or obsolete items. Ever more materials are required to make the products, and because the original manufacturer does not have a responsibility for taking back products after the end of the lifetime, the users discard the disused products as waste.

Sarah encountered three dominant negative linear design incentives that are very common in the cycling industry.

1. Products are designed to fail

In many cases, companies reduce the useful life of a product deliberately, in order to be able to sell more products and sell them faster. Probably the best-known example of such 'planned obsolescence' is the lightbulb, which technically can survive for many decades. The industry recognised long-lasting products were bad for business and set up the 1,000 Hour Life Committee in 1924 to make sure each manufacturer produced lightbulbs with a maximum of 1,000 hours of light before they broke down.[11]

Bike parts fail because, for example, too little material was used in critical places. Often this concept is justified with the argument that a product needs to be as light as possible. An example is a saddle, which hardly ever outlasts a frameset because it cracks at the bottom of the shell. A little extra material would make them last forever. Instead, the entire saddle is thrown away, and a new saddle is picked up at the shop.

2. Products are designed to be outdated

This is very common in the computer and mobile phone industry. Every few years, a new model is introduced with more computer power and new interfaces between the device and, for example, a charger. The new interface requires new cables, making the old cables instantly worthless. Software updates are even worse culprits, as they are often also pushed to older generations of the product with less memory and computing power. The updates slow down the older models and reduce their battery life, forcing the consumer to buy a new one.

New generations of parts in the cycling industry cause the same problem. Backwards compatibility is not part of the general design principles. Instead, the change from Shimano 10-speed to 11-speed and, more recently, the introduction of the SRAM XDR™ 12-speed cassette bodies have made many older generation wheelsets useless. The same goes for the rest of the drivetrain: shifters, derailleurs, chains and cassettes become waste if one of the parts in the groupset fails and is no longer available as a replacement part.

3. Products are designed to be out-fashioned

Before the industrial revolution, most goods were made to order. Someone who needed, say, a bicycle, would go to a blacksmith and order a new one. Marketing was simple: if you were a producer, people had to know where you were and what you could do for them. Mass production changed the way producers and consumers interacted completely. All of a sudden, products were made without a guaranteed customer and marketeers started to look for new ways to sell the 'overproduction'. Marketing methods have been perfected ever since, making us buy all sorts of items we never thought we needed.

If you compare bike ads over the last couple of decades, you can clearly see how this works. Almost every brand uses exactly the same words in every single ad every year: faster, stiffer, lighter, better, more comfortable. The tiniest little improvement is presented as a breakthrough and makes most cyclists long for their next big purchase. Once a new bike shows up on a weekly group ride, all the other cyclists take a close look, compliment the owner, and start to think that they too require the latest technology to improve their riding.

2.3 LIMITS TO GROWTH

You might wonder what the fuss is all about, this talk about the downsides of the linear economy. The problem is that with an exponential increase in world population and wealth (see Figure 2), our environment suffers from a pattern of rapid increase of resource extraction, waste production and pollution. Before the industrial revolution, the number of people living on our planet could be supported by the natural systems around us, absorbing the CO_2 emissions from our wood fires, growing new trees to replace the ones we needed to build our houses, and restoring stocks of fish after a successful fishing trip. Now, all of these 'common resource pools' are over-exploited by countries and companies to generate wealth in the short term.

In 2018, the World Economic Forum published its 13th Global Risk Report that describes the main challenges the global economy is facing. Risks related to the environment have grown in importance recently, and the most severe among them are directly related to climate change:[12]

1. 'Extreme weather events' such as floods and storms causing damage to infrastructure and property and loss of life;
2. 'Failure of climate-change mitigation and adaptation by governments and businesses' to protect populations and businesses from the impact of climate change.

177 Gt
MASS
Material extraction
in billion tons (Gt)

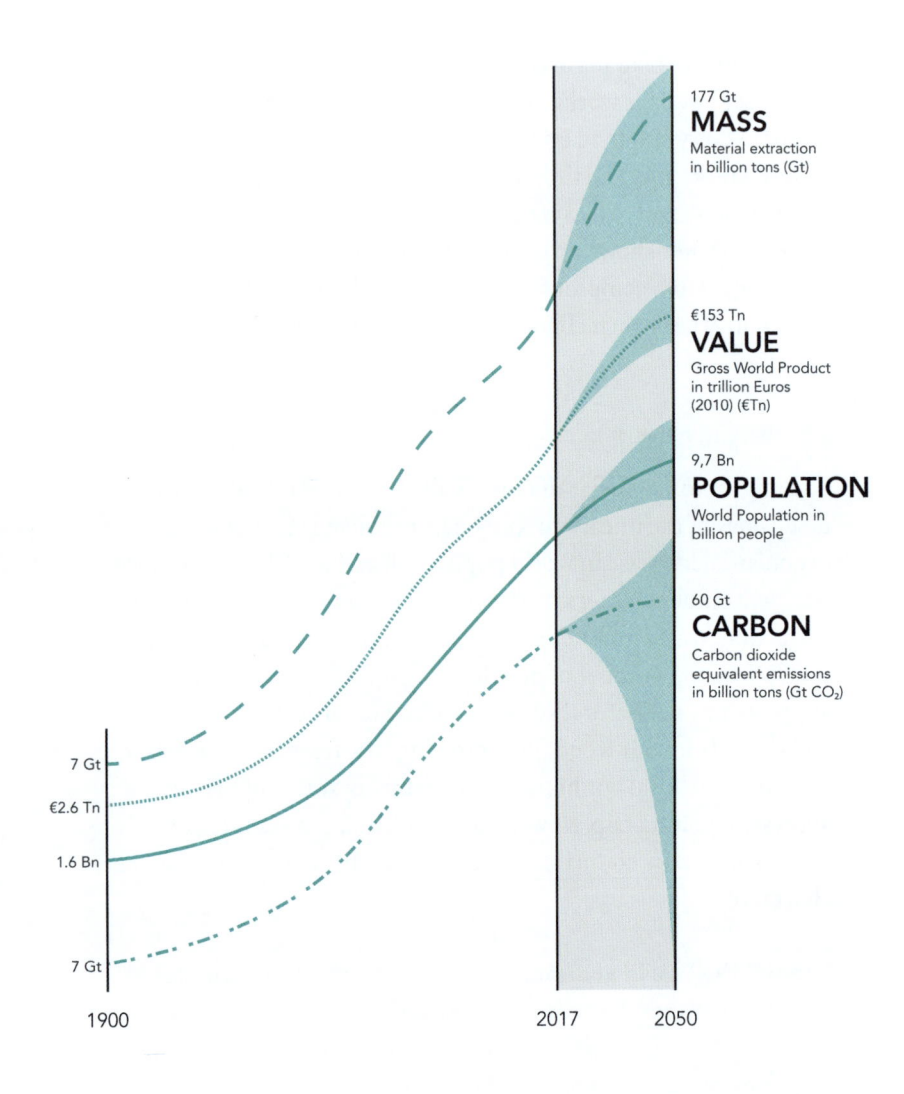

€153 Tn
VALUE
Gross World Product
in trillion Euros
(2010) (€Tn)

9,7 Bn
POPULATION
World Population in
billion people

60 Gt
CARBON
Carbon dioxide
equivalent emissions
in billion tons (Gt CO$_2$)

7 Gt
€2.6 Tn
1.6 Bn
7 Gt

1900   2017   2050

Figure 2: Development and projection of material extraction, gross world product, population and carbon dioxide emissions between 1900 and 2050.

If we want a 50% chance to prevent the earth from warming more than 1.5ºC above pre-industrial levels, as agreed in Paris in 2006, we can continue emitting CO2 emissions from fossil fuels in the amounts we do today only until mid-2031.[13]

Resource extraction is another environmental and economic risk. The average European uses about 14,000kg of resources a year, including indirect materials required for the extraction of resources, losses in manufacturing processes, energy production, and for building infrastructure.[14] In total, about 92.1 billion metric tons of materials are extracted from the earth, and only 8 billion metric tons of materials are recycled. This means that the world is currently only about 8.6% circular.[15]

Some materials are widely available now, but could become much harder to extract even by the end of this decade. Cobalt, a rare earth metal, is one of the critical materials in battery production of which we could see shortages within just a few years, especially when you consider that half of the supply is coming from the Democratic Republic of the Congo. Reports of human rights abuses and environmental impact due to hazardous materials (up to 2,000 metric tons of waste are created for the mining of a single metric ton of rare earth metals) are common for this country, which contributes to the risk of supply disruption as well as reputational damage.[16] Understanding the supply chain of a product and its dependency on rare materials is important to prevent supply shortages. The Dutch government launched a 'Raw Material Scanner' to help companies understanding the risks.[17]

The exploitation of these global commons creates global wealth, but at the same time has a worldwide negative impact on our environment. This problem is difficult to solve, because we do not have a global government to decide by whom, when and in what quantities the commons can be used, meaning that action is needed by governments, companies, NGOs and citizens to collectively manage the commons responsibly.

RUBBER SCARCITY CALLS FOR ALTERNATIVES: RUSSIAN DANDELIONS & INNER TUBE RECYCLING

Natural rubber is probably the only biobased material source used in a road bike today. It is a rare example of a material with better performance than non-bio alternatives, as natural rubber has better elasticity and energy absorbing properties than the synthetic alternative, which makes it the ideal material for tyres. The rubber is tapped from rubber trees in Thailand and Indonesia mainly, where temperature, humidity and rainfall conditions are optimal. However, the increase in production of natural rubber is nearing its limits. Grown in large monoculture plantations, the trees are vulnerable to diseases; the further increase of the size of the plantations requires the destruction of tropical rainforest, which is no longer considered a viable option. This means that the supply cannot keep up with the explosive growth in demand for rubber for tyres (mainly for cars) and other products. Without the development of alternatives, natural rubber demand is expected to exceed supply by around 2022/2023.[18]

To deal with this potential scarcity, several large tyre manufacturers have introduced tyres with rubber from an alternative source: the roots of Russian dandelions (Taraxacum koksaghyz). The dandelions can be grown in places in Europe and other areas with similar conditions, where they need just a year to grow before latex can be harvested from the roots. Although the supply of dandelion rubber is still limited, Apollo Vredestein has launched their first batch of limited edition 25mm Fortezza Flower Power high-end road tyres; Continental offers their own limited edition 35mm Urban Taraxagum™ urban tyre.

Another manufacturer, Schwalbe, collects synthetic butyl inner tubes at dealers to send them back to their factory in Jakarta, where they are recycled into new tubes. This process, including the transport from dealers in Europe to the factory in Indonesia, requires about one sixth of the energy compared to the production of new butyl. To put this into perspective, it requires about 1 hour of cycling at 200 watts to produce a single 65 gram extra light tube.

Some 20% of butyl used for the production of tubes in this factory currently comes from recycled sources. All you need to do to increase this percentage is to drop your punctured tube off at your local bike shop, so they can return the rubber to Schwalbe.[19]

Litres of fresh water consumed per kilogram of product manufactured.

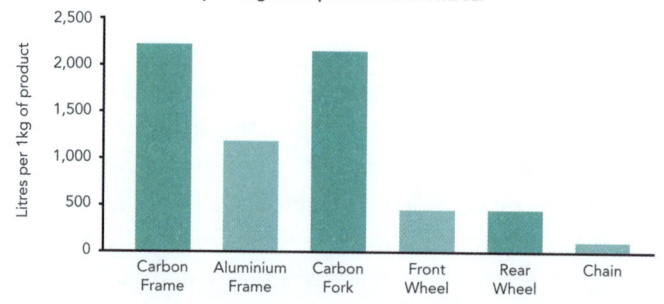

Embedded energy in gross caloric value (kWh) per one kilogram of product manufactured.

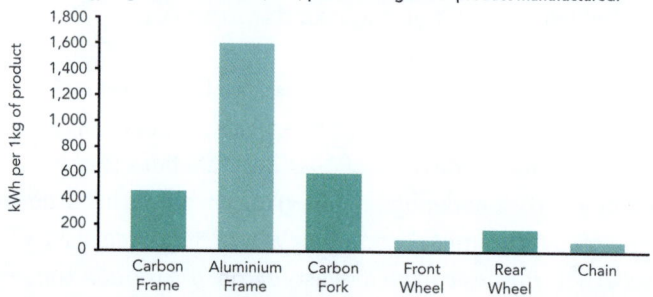

Global warming potential in kilograms of CO_2 equivalents per one kilogram of product produced.

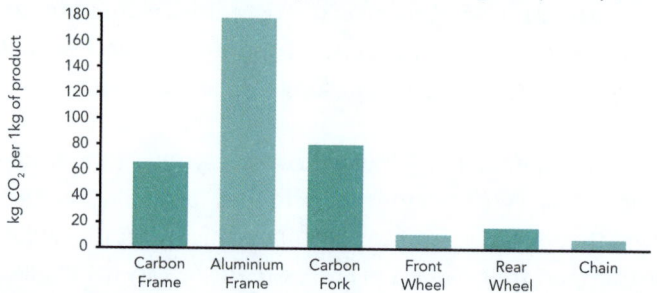

Solid waste in kilograms per one kilogram of product produced.

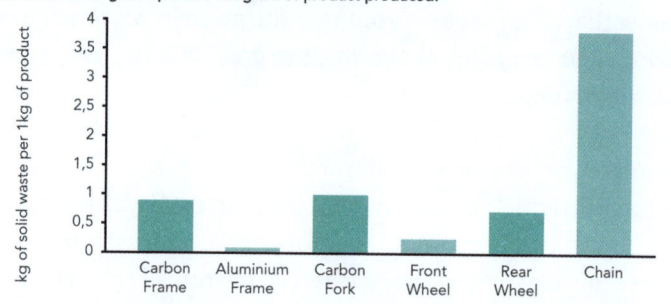

Figure 3: Results of Specialized LCA Study: Water use, Embedded energy, Carbon Dioxide emissions and Solid waste per kg of product produced.

2.4 MEASURING ENVIRONMENTAL IMPACT

Effectively improving the environmental impact of products starts by understanding where the biggest impact can be achieved with the least effort. Companies can adopt global standards to measure CO_2 emissions using the Greenhouse Gas Protocol.[20] This helps to find out where the biggest improvement for the short term can be made. This includes emissions related to electricity use, office heating and employee travel.

To understand the complete environmental impact of a product, you have to consider the entire supply chain. A standardised Life Cycle Analysis (LCA) method has been developed to quantify impact on several topics, such as biodiversity loss, water and energy loss, toxicity and global warming (through the emissions of greenhouse gases such as CO_2). The analysis includes production processes, distribution, use and end-of-life scenarios of a product.

In 2013, Specialized Bicycle Components commissioned Duke University to conduct an LCA study that includes the analysis of four types of impact (fresh water consumption, embedded energy, global warming and solid waste) of a carbon fibre Specialized Roubaix and an aluminium Specialized Allez frameset, a carbon fork, a DT wheelset and a SRAM chain.[21] The LCA has limited accuracy, as most of the suppliers contacted by the research team could not or would not supply the required data, but it gives a good indication of the difference in impact for each product category and where there is potential for improvement. The graphs in Figure 3 show the difference between the products, all shown per kg of product and not in weight of the actual product, to make it easier to make a comparison.

According to the study, the production of a kilogram of carbon fibre requires 2,160 litres of fresh water compared to 1,240 litres for a kg of the aluminium frame. When it comes to energy use and the impact on the climate, the aluminium frame scores worst. The manufacturing of a kg of the aluminium frame requires some 1,600 kWh of energy, more than 50% of the electricity an average Dutch household uses in an entire year.[22] The corresponding CO_2 emissions caused by this energy use equals over 170kg for a single kilogram of aluminium frame!

Carbon frame manufacturing generates much more solid waste than aluminium frame manufacturing: about one kg of solid waste to one kg of frame. The production of the chain scores even worse, close to four kg of waste to one kg of chain. Manufacturing a 259-gram steel SRAM PC 1071 chain therefore generates 955 grams of production waste. Even though chains show bad results for total waste, at least the metal waste from chain manufacturing is recycled, whereas most of the waste in the carbon fibre frame manufacturing process cannot be recycled and will end up in a landfill or an incinerator. The study, although not perfect, provides good insight into differences in materials and production processes, and shows you where to focus.

ENVIRONMENTAL PRODUCT DECLARATION (EPD)

The globally standardised way to measure and report the environmental impact of products is through an Environmental Product Declaration (EPD). Its main goal is to quantify the environmental impact of different products, and hence make it possible to compare them. The EPD is standardised by ISO 14025, which was first published in 2000, and revised and updated in 2006. The product's Life Cycle Analysis (LCA; ISO 14040) plays an important role in the EPD.

Every sector has its own specific properties, which are included in so-called Product Category Rules (PCRs), which guide the EPD measuring and reporting process and guarantee comparability within a sector. To date, there are no PCRs defined for the cycling industry. Introducing the EPD in the cycling industry would help enormously to raise consciousness about the role of different players within the industry, to identify critical rare materials, and to measure and control the environmental impact of the entire supply chain.[23]

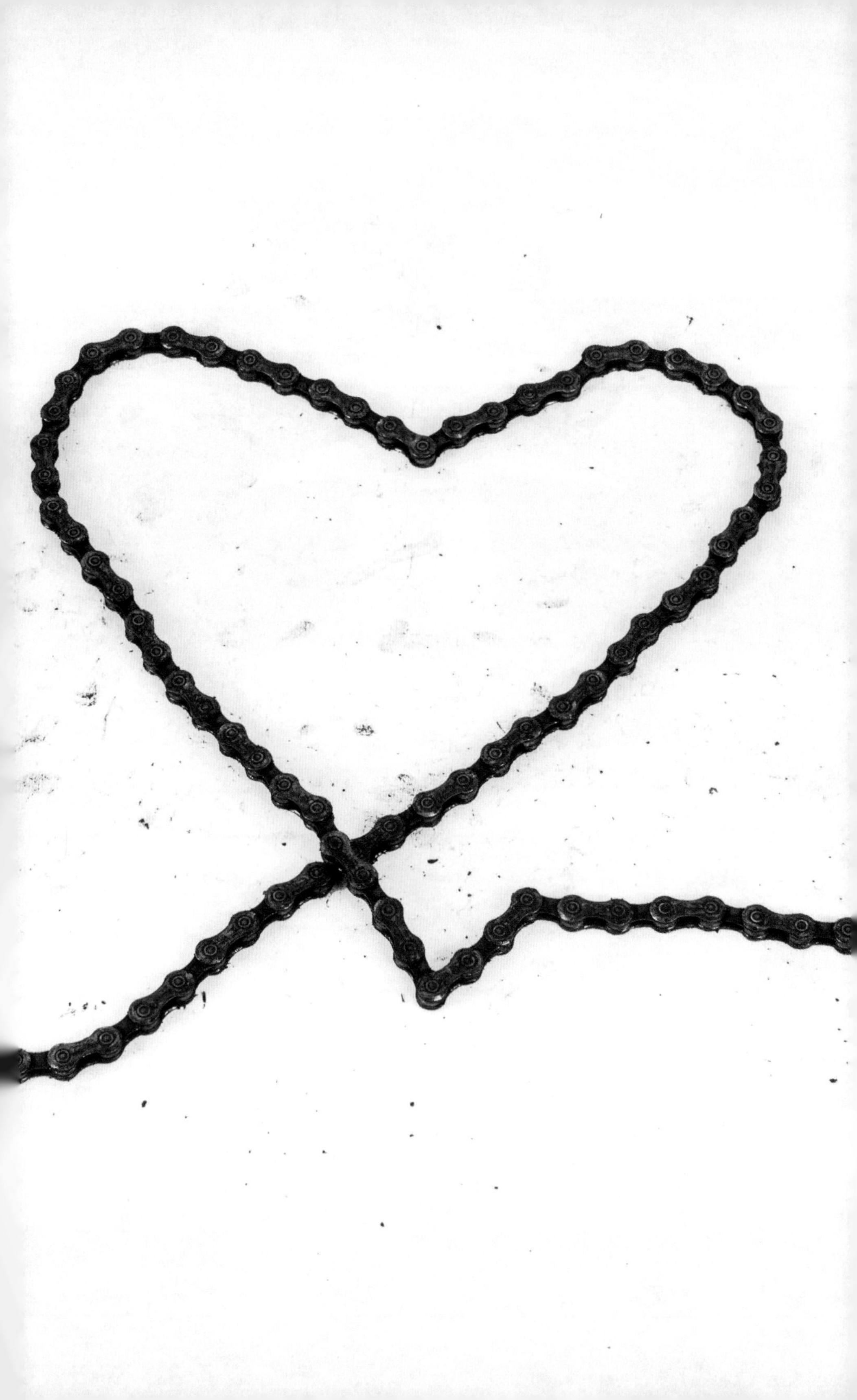

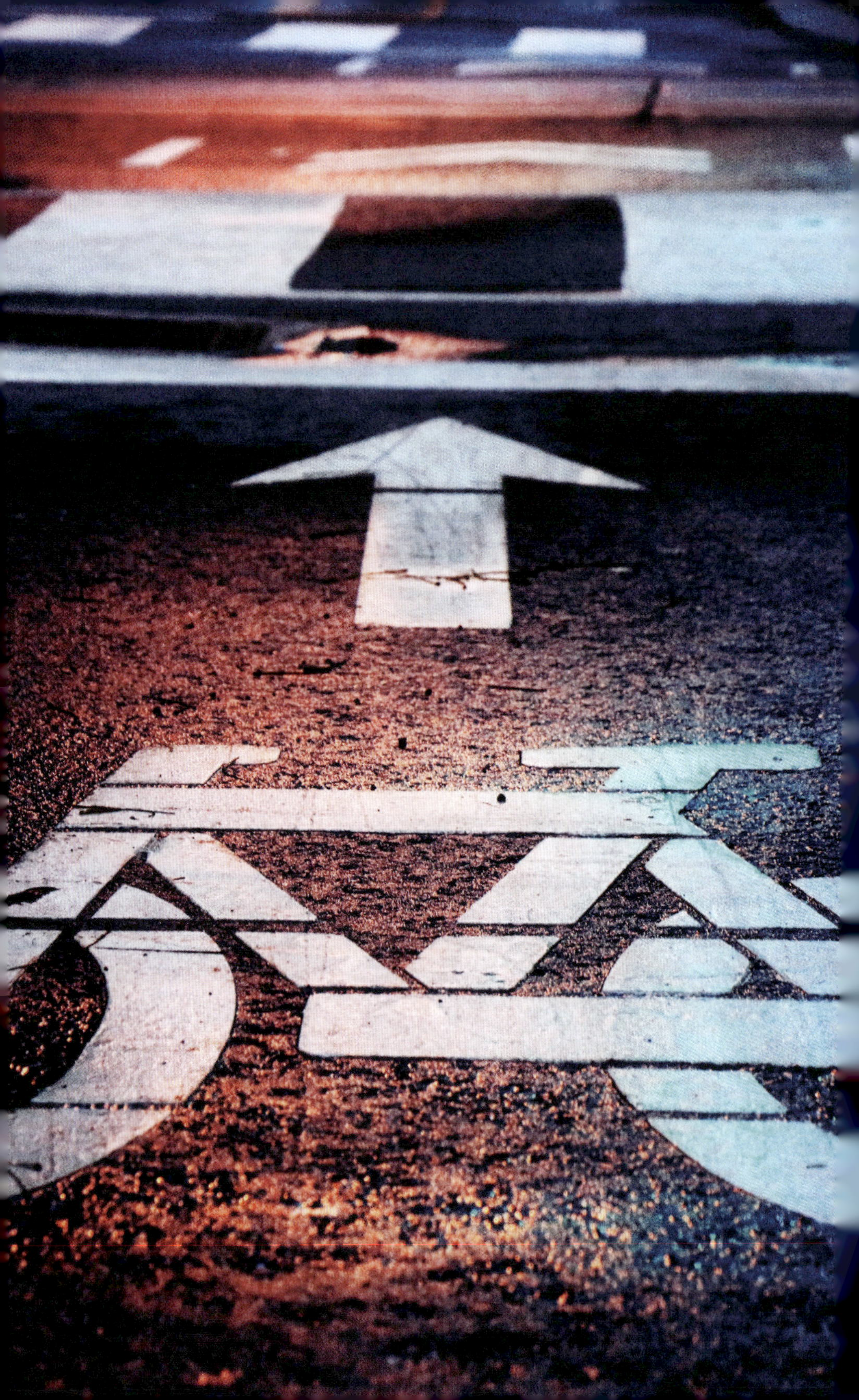

STAGE 3
CREATING VALUE IN A CIRCULAR CYCLING ECONOMY

In a circular economy, companies make money by offering products to customers ensuring that materials constantly flow from resource to product, and back to resource, without producing waste and without causing pollution in any form, whether it be solid, liquid or emissions to the air. It is a 'closed-loop system' where nothing is wasted. It requires us to think differently about our business models, our design philosophy and the materials that we choose.

You can do a quick 'mental check' to see whether a product fits in a circular economy:

- Do the production and transportation of this product require finite resources and/or fossil fuels?
- Does the product contain hazardous (toxic) chemicals, or have hazardous chemicals been used during the production process?
- Do contaminant (toxic or other non-biodegradable) materials and emissions end up in the environment during the use of the product?
- Do any materials, which together form a product, end up as waste in nature, in a landfill or an incinerator, or in a recycling process?

If any of the above questions are answered with 'I don't know', 'yes' or 'partly' you have work to do. The first step in the transition from a linear to a circular economy is to identify where money can be made through a reduction in cost or an increase in revenue. Chances for cost savings or new business can be found in parts of the linear economy producing both economic and ecological waste. The challenge is to close the loop through the development of new designs and new business models addressing this combination of economic and ecological waste.

There are four reasons why companies should act now:

1. Brand continuity requires innovation, and innovation requires focus and creativity. Focus arises from the need to develop circular products, while creativity comes automatically with such a radical new direction in product design.

2. Consumers' attitudes towards brands are changing. More and more consumers are looking for sustainable alternatives; the market for organic food is growing fast and there is hardly a big outdoor clothing brand left without a proper sustainability strategy, because consumers are inclined to pick the more sustainable of the available products. Inaction will result in the loss of market share to other brands that will start offering products with a lower environmental impact or, even better, with environmental benefits.

3. Governments are already changing regulations, such as fuel economy standards for cars and carbon tax systems, favouring products with smaller environmental footprints and less waste, in line with long term ambitions, such as the EU's vision for 2050.

4. If we fail to make the transition towards a less resource-intensive production system, we will reach a point where material sources will be depleted and materials will become more expensive.

LINEAR ECONOMY

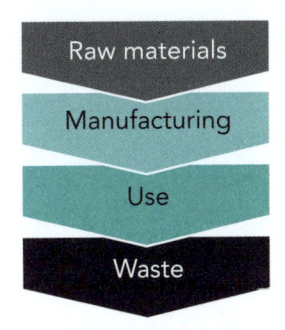

Raw materials
Manufacturing
Use
Waste

CHAIN ECONOMY WITH RECYCLING

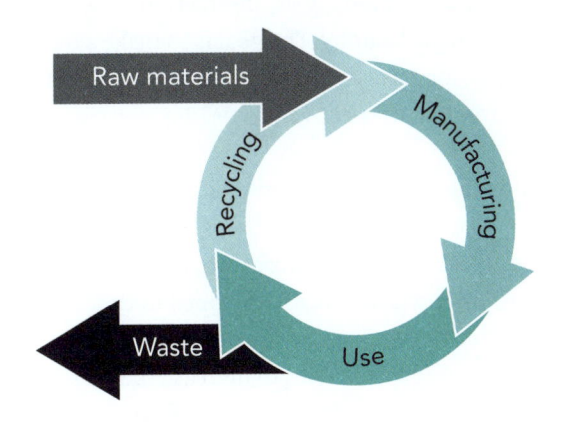

Raw materials
Manufacturing
Use
Recycling
Waste

CIRCULAR ECONOMY

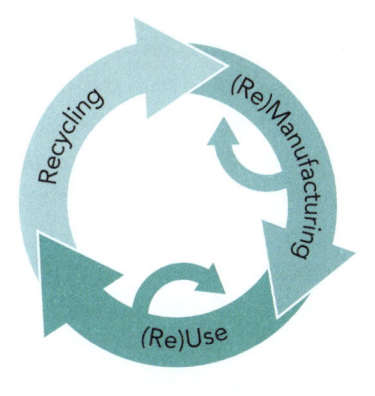

(Re)Manufacturing
(Re)Use
Recycling

Figure 4: A linear economy, a recycling economy and a circular economy.

3.1 FROM LINEAR TO CIRCULAR – WHERE ARE THE OPPORTUNITIES?

We believe that the best way to make the transition from a linear to a circular economy is through business, not through waiting for government regulation or, worse, resource depletion or natural disasters caused by a changing climate. If it makes economic sense to improve the design of a product or the interaction with a user, resulting in ecological benefits, however small, it is a step in the right direction.

There are numerous organisations all over the world that have been contributing to the concept of a circular economy. We have used ideas described by, for example, the Ellen Macarthur Foundation[24], Circle Economy[25], Products that Last[26] and CIRCO[27]. They all describe how the core of a circular economy is the interaction between a business model and a design philosophy. The goal is to make sure that a product is used as long as possible, optimised from both an economic and an environmental perspective.

This means that the value that is created in the pre-use phase should be retained in the use and post-use phases at the highest level for as long as possible. This often requires smart design features and rethinking customer relationships developed in the pre-use phase of a product. Different business models and types of products require different circular design strategies to increase the lifetime of a product and reduce waste at the end of a product's useful life.

Figure 5 gives an idea of how economic value develops over the lifetime of a bicycle in the linear economy. We took a road bike with a suggested retail price of €3,000 as an example and made an estimation of the value attached to each step of the linear lifecycle of a product. To make a €3,000 bike, you need to start with buying raw materials, which cost about €30, depending on the type of material and the material losses during the manufacturing process. Machining and other types of processing raise the value of the materials to about €100. Once assembled into parts and then a bike, the bike is sold to a distributor for €1,300. Distribution from the factory to the shop, marketing and taxes further increase the price of the bike to the maximum value. If the bike is not sold in the year it was launched, it will be sold in the annual end-of-season sale with a 25% discount.

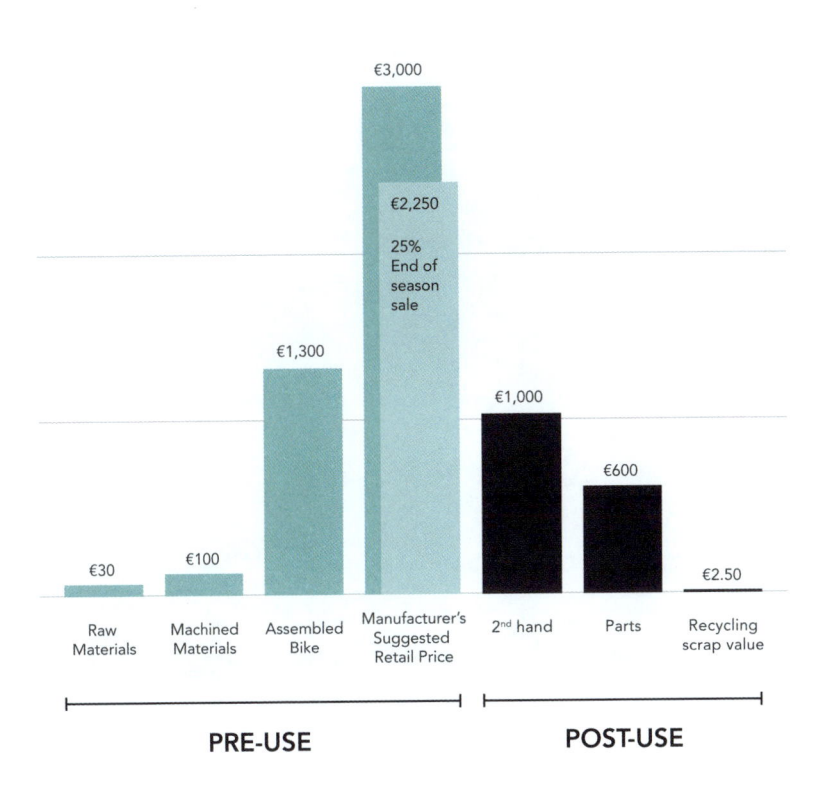

Figure 5: Economic Value over the lifecycle of a €3,000 road bike.

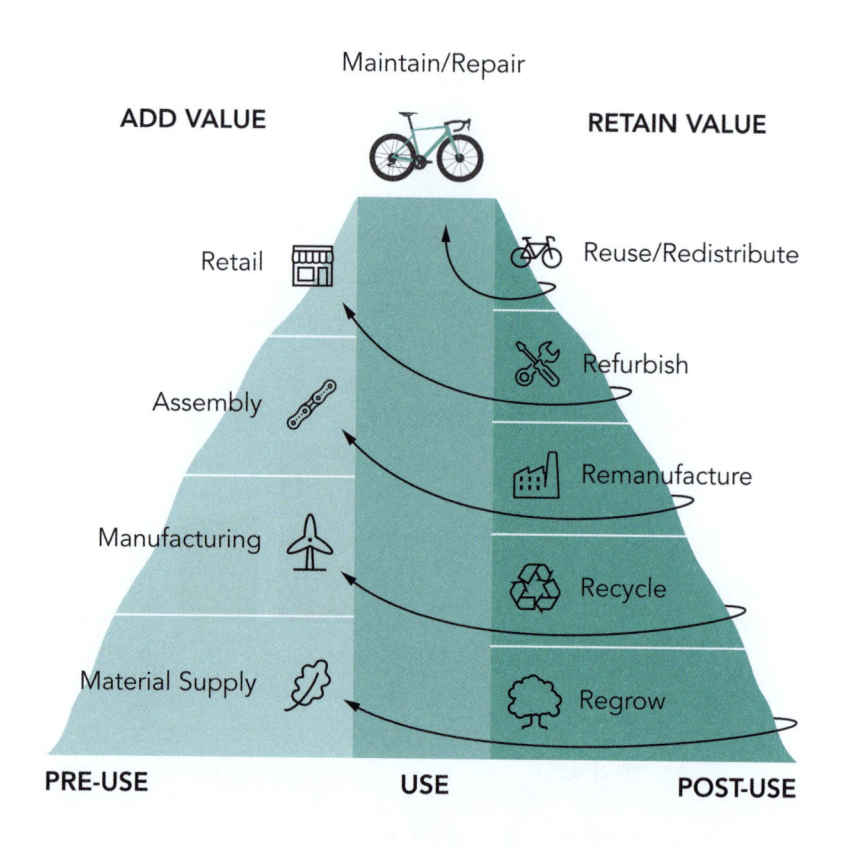

Figure 6: Circular cycling economy Value Hill.

After the customer no longer uses the product, it is often sold at about a third of the original price to someone else, who will reuse the bike. After some time, the bike is considered worn, and the parts could be sold individually and bring in about €600 if the owner were to go to the trouble of disassembling the bike and selling the parts separately. Once the parts no longer yield any value on the second-hand market and they are thrown away, the recycling scrap value of the materials is just €2.50 – assuming the materials can be separated in the local recycling process in the first place.

This summary makes it clear just how much value is added to materials in the process of producing a bike, and how quickly it is lost again. The value of the scrap materials is even a lot less than the value of the original materials, because a lot of effort is required in the processes of transportation, separation and actual recycling of the materials before they are good enough for another lifecycle.

In Stage 2 we presented a customised Value Hill for the linear bicycle lifecycle. **The Value Hill for a circular product (Figure 6) looks similar, but there are four important differences:**

1. The use phase is much longer because the products last longer and are easier to repair.
2. There is no waste. Products and materials are fed back from the post-use phase into the pre-use phase at the highest possible value level through reuse/redistribution, refurbishment, remanufacturing, recycling or regrowth of materials. Materials are no longer sourced from finite sources, but come from the post-use phase of other products or from renewable sources.
3. There is no pollution anywhere in the lifecycle.
4. Renewable materials are harvested from natural resources that regenerate ecosystems by increasing biodiversity and resilience to the changing climate.

Design has a huge impact on the entire lifetime of a product, and on pollution and waste as a result of the production, use and end-of-life scenario of a product. Therefore, a designer needs to take a completely different perspective on product design: instead of designing a product that fits the short-term fashion trends, in line with the linear economy, the challenge for product designers in a circular economy is to look at the product taking the end of its lifetime as a start. If a product is designed with a clear end-of-life view aimed at preventing waste and pollution during the entire lifecycle of a product, the design focus will automatically shift towards retaining value through longer use, reuse, refurbishment and remanufacturing of a product. And if used products and renewable (biobased) material sources become part of our production processes, we will automatically be able to reduce the use of finite materials and start regenerating the environment through the need to create new sources of renewable materials.

There are several ways to retain the value of materials. Prolonging the lifespan of a product starts at the top of the hill through maintenance and repair. Every step the product descends down the hill reduces its value, as more time and energy will need to be invested to prepare the material for another lifecycle. We have identified six strategies to prevent materials from turning into waste. They all require some sort of reverse logistics to bring parts back to a place where their value is retained.

I. MAINTAIN/REPAIR

Maintain/Repair is the first strategy to keep the value of a product at the highest possible level, prolonging the life of a product, often at the lowest cost.

Traditionally, the cycling industry has had a flourishing maintain and repair ecosystem. Bike shops in even the smallest villages have been repairing and replacing parts for many decades. Many cyclists know how to repair and replace parts on a bike at home with simple tools, keeping bikes on the road for many years.

2. REUSE/REDISTRIBUTE

The value of a used bike has been recognised for a very long time and retaining value through this strategy is very common in the cycling industry. There is a thriving second-hand market for parts and bikes on websites such as eBay, where bikes are sold at very attractive prices, sometimes 75% under the original sales price in the shop. Many deals are made online, and parts are sent to buyers in other parts of a country or even on the other side of the planet.

Second-hand bikes typically are not completely checked or even properly cleaned before they are sold, to keep the costs down for the selling party. Buying a second-hand bike or part always implies the risk that parts do not have the quality or remaining lifetime of a new part, which justifies a lower price, to allow the buyer to pay for potential future replacements. Bike shops use this business opportunity by selling second-hand bikes with, for example, a one-year warranty at a price between the consumer-to-consumer deals on the internet and the price of a new bike.

3. REFURBISH

Refurbishment is the process of restoring a product to a quality level close to that of a new product, through repair and replacement of certain parts. A refurbished product is often made from a mix of used and new parts, and sold with a warranty provided by the refurbishment company to a new owner.

To be able to guarantee quality, a product is often completely disassembled to be able to check individual parts and replace broken or worn parts. Refurbished products are sold at prices between reused and new products.

THE STORY OF CIRCULAR CYCLING

In 2018, we opened Circular Cycling, a small shop in the city of Utrecht, right in the heart of the Netherlands. We made and sold refurbished road bikes that we dubbed UpCycles. We built new bikes from used parts to test how circular business models could work for the cycling industry.

We had already spent a couple of years working out how best to start a business with the limited resources that we had available. We combined our experience of sustainability and circular economy in the construction industry, and inside knowledge from the cycling industry, to figure out what would work best. We read many books and websites, took CIRCO circular business model training, and entered the EU-funded Climate-KIC Accelerator programme for sustainable start-ups.

We soon realised that if you want to enter the road bike market with a new type of product, you need to focus on a specific market segment. We decided not to target the 'bike freaks' (who turned out to be our best suppliers) and competitive cyclists, because both groups highly value the latest technology. Without them, the market is still huge – most people ride their bikes for fun without an excessive interest in the latest technology. We also found that more and more cyclists are concerned about reliable and sustainable products. In order to cater for this market, our bikes needed a warranty equal to one for a new bike (two years) and an attractive price to compete with new bikes of around €1,500. And the bikes needed to come with a good story, which the buyers could share during their next group ride.

The 'resources' for the UpCycles were frames and parts that were given to us or we bought from other cyclists, shops and distributors. They all had a Box with parts which were too good to throw away but were no longer used. Some parts were new, others used but in good condition, some were beyond repair. Being able to purchase the parts at a low price was an important part of the business case, and we knew from the beginning that the challenge was to keep labour costs down to be able to make a profit.

We took all the bikes that we got completely apart, so we could check the separate components for quality. Taking a bike apart is done in no time and has two advantages: it makes it easier to do quality control and, just as important, it allowed us to mix and match parts from various sources to customise each and every bike we put together. Each bike we assembled should be of such high quality that we would be proud to ride it ourselves. We even gave each one a unique name, and thereby an identity, to improve the emotional bond between the rider and the UpCycle. Bikes like Addiction, Angel and Yellow Monster were all given an online Bike Passport (more about this later) containing photos and a detailed description of all the parts used on the bike, as well as the remaining expected lifetime.

To guarantee ride quality and safety, every UpCycle was fitted with modern top end tyres, new brake/shift cables, a new compact handlebar and new bar tape. All other parts were only replaced with new parts if we did not have pre-used alternatives with sufficient remaining lifetime available.

Yellow Monster
Cannondale R2000si
UpCycle

circular·cycling

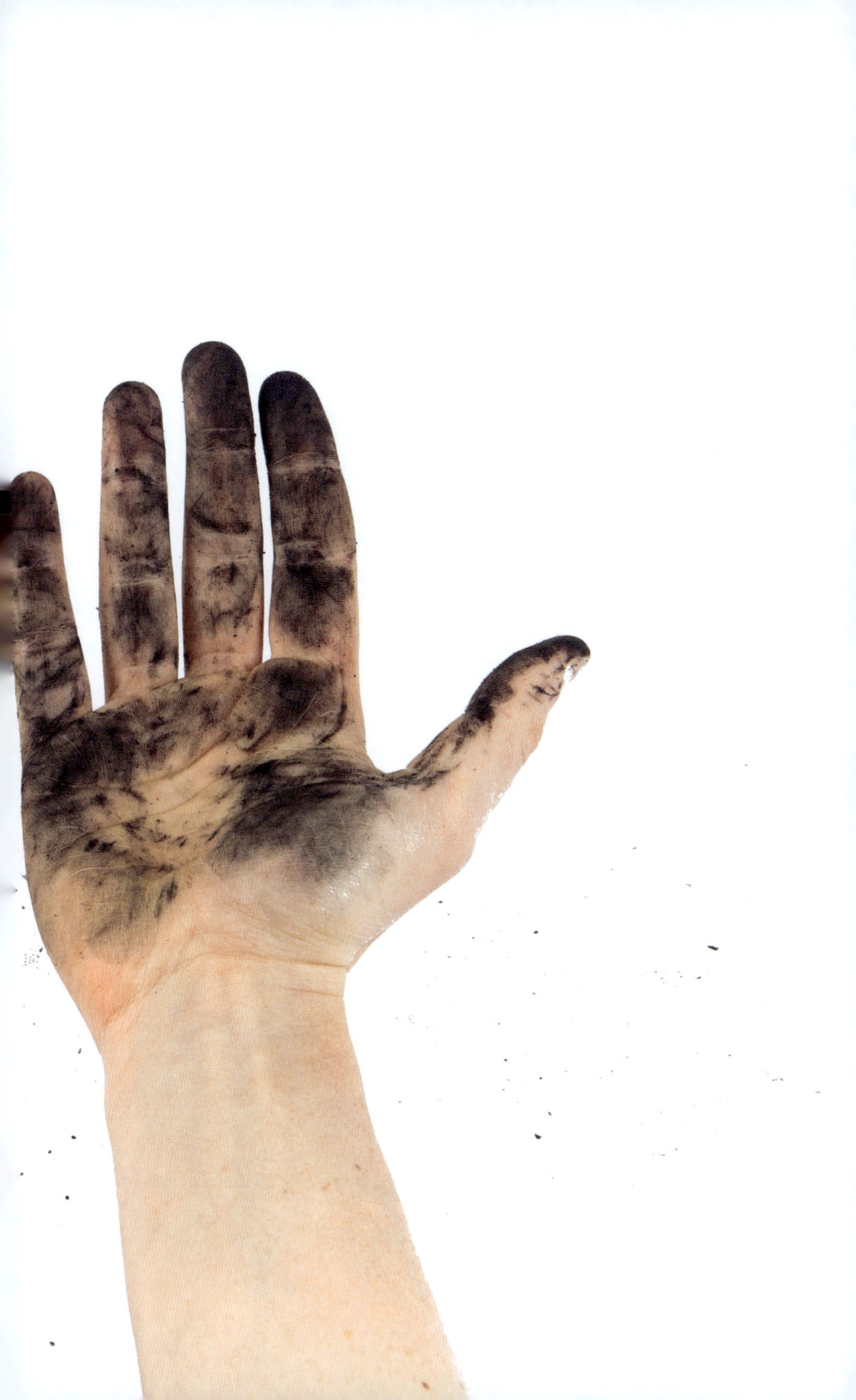

Every bike that left the shop felt like new. Especially the former top-range models were excellent bikes for the price of a mid-range new bike; they put a smile on the face of every bike connoisseur we showed them to.

As with any start-up, it took some trial and error to figure out what worked and what did not. Eventually, we decided to stop building the UpCycles. The whole experience, including its upsides and downsides, is worth sharing, because it highlights some of the limitations of using linear products for circular business models.

What went well:

1. Bikes and parts from model year 2005 and onwards were very suitable for refurbishment. Quality was high, compact cranks in combination with 11-28 cassettes offered suitable gear ratios, and combined with modern tyres and ergonomic handlebars, they looked and felt like new.

2. Many parts that we recovered from Boxes had a considerable lifetime left; after a good clean they were very suitable for use on refurbished bikes. On average, we were able to build an UpCycle from about 90% reused materials.

3. We got very positive press coverage in traditional media and many sign-ups to our e-mail newsletter once our concept was picked up by the press, both within and outside the cycling industry.

4. Most of our customers specifically chose our UpCycles over a new bike, because they wanted an eco-friendly alternative.

5. We did not have a single warranty issue.

6. We learnt a lot, met many great people and had loads of fun.

What did not go so well:
1. It took too long to build a complete UpCycle

We founded the company based on the idea that bikes are good examples of products that are easy to disassemble and reassemble, parts being interchangeable between different frames, and spare parts being widely available. Although this sounds obvious, it gets harder and harder to maintain a bike's value through maintenance, reuse and refurbishment:

- There is increasing variety in 'standards' for parts such as bottom brackets, wheel/frame interfaces, cassette bodies, shifters and brake pads. We found that reliable information about parts and their compatibility was hard or even impossible to find, even on websites of manufacturers who did make an effort to provide this kind of information. This made it difficult to identify which parts were compatible.

- If a part needs replacement, the availability of specific parts on the second-hand market, in shops and even at distributors is very limited, or even non-existent. Trying to get hold of these parts proved, again, very time consuming.

- The integration of cables for brakes and gears into frames and handlebars makes a bike look cleaner and more aero as long as you ride at the speed of a pro rider, but does not really help to reduce the time to build or maintain a bike. We found that this extra hassle means that simple repair jobs will sometimes take twice the time (and a lot of cursing), even with proper tools and tactics.

- It seems that with every new standard, a new set of specific tools is required, often more expensive than the part it is meant for. We found that this was a barrier for a small bike repair shop like ours. It also means that a mechanic needs to spend time on determining which tool to use and on finding it. This could provoke the use of improper tools for the job, with parts being damaged as a result.

2. Marketing

When you launch a new product in a market as a start-up company with limited funds, reaching customers is always going to be difficult. We found that online advertising for a new market segment like refurbished bikes is complicated. Hardly any people use keywords such as 'sustainable bike' or 'refurbished bike' on online search engines. It was easy to become the top result for those searches, but they hardly generated additional traffic to our website. We tested targeting 'second hand' and 'discount bike' searches, which did generate a lot more traffic, but very little turnover because our UpCycles turned out not to be what people were looking for: they were either too expensive for a second-hand bike or not part of last season's sale. Compared to big brands, we had very limited marketing budgets and no social media channels with large numbers of followers, which made it really hard to quickly generate awareness of our product.

3. The impact of design trends on compatibility

When we reached a point where we had to decide to make a large investment to scale up the company, we reviewed the assumptions in our initial business plan. An important industry trend that developed faster and more widely than we anticipated, was the increased integration of parts into complete 'packages' with special stems, seatposts and integrated cables, the shift to disc brakes with all the different axle designs, and finally new 12-speed compatible freewheels. Although we were confident that these details would not matter too much for our targeted customer group, they make it very hard to run a refurbishment business based on the idea that bikes were built with interchangeable parts. We estimated the required size of our future inventory, and concluded that we would need an incredible number of Boxes...

4. REMANUFACTURE

The step from refurbishment to remanufacturing is made at the level of parts and materials. When remanufacturing parts, used parts and materials are fed back into a production system as part of a mix with new materials at the original manufacturer. This reduces the need for new resources by (partially) replacing resources with parts and materials from used products.

A simple example of remanufacturing in the cycling industry could be to feed used seatposts (a typical product with a significantly longer life span than the average part of a bike) back into a production line. They would have to go through a quality control procedure to check their integrity, and then be re-anodised to make them look as new before being sold as brand-new seatposts.

5. RECYCLE

Recycling sounds nice and feels good, right? Don't be fooled. Recycling often involves a lot of transport, and a lot of energy to separate and recycle materials. And even after various ingenious processes have separated materials that are bonded together, a lot of material that is considered worthless is, for example, burnt in the process that melts scrap metal back to new metal. Most materials that are considered 'recycled' will not return to their original value and can be seen as downcycled to a lower value.

In a circular economy, recycling is the path of last resort, the final step before material that cannot be returned to the biosphere ends up in a landfill or an incinerator. For a long time, 'but we use recyclable material' was the answer when companies were asked about sustainability and resource use. There are multiple reasons why recycling is not as effective compared to the actions higher up on the Value Hill. Take a complete bike for example – if someone wants to dispose a bike, he/she can put it in the garbage or, slightly better, take it to a recycling station. There, the owner needs to make a choice: does my carbon fibre bike go in the container labelled 'metal', 'plastic' or 'other'? It is a tough question without a clear answer, because to properly recycle a bicycle, it has to be taken apart completely. Even if you decided to go through the hassle of that, it is impossible to separate an aluminium crankarm with a carbon fibre wrap or a rear derailleur made from a mix of carbon fibre, steel, aluminium and a number of other materials. What will probably happen, is that the bike will end up in the 'metal' container at the local recycling station. From there the bike will be sold and transported as scrap metal to a recycling facility anywhere on the planet. Once there, the bike will be shredded into small pieces of ferrous and non-ferrous metals, which will then be separated and be fed back into the steel making process. All remaining non-metal parts will either end up on a landfill or be burnt and lost forever.

RECYCLING CARBON FIBRE, THE MIRACLE MATERIAL

Since the introduction of carbon fibre in the cycling industry several decades ago, this 'miracle material' has slowly surpassed steel and aluminium as the material of choice for many parts of a bicycle. Its unique combination of high stiffness, low weight and freedom of form makes it a great material for frames, seatposts, cranks and even rims.

The composite material gets its strength from long carbon fibres that are bonded together with a sort of glue, called epoxy. Once the epoxy has cured, the fibres offer strength in all directions.

Apart from the health issues related to carbon fibre dust ending up in people's lungs when they work with the material during production, the problem with carbon fibre is that it cannot be melted like steel and aluminium. Instead, the epoxy has to be burnt off or dissolved in chemicals to get the valuable carbon fibres out again. According to a Life Cycle Analysis done by the carbon fibre recycling industry, this process requires only 10% of the energy required for the production of new fibres.[28] However, the fibres resulting from the recycling process are shorter than the original fibres and therefore have different properties. In addition, there is a need to add new epoxy to bind them together. In other words: the fibres are downcycled.

To turn carbon fibre recycling into an efficient and cost-effective process, three things are required: sufficient supply of discarded carbon fibre, a large enough market for recycled fibres and a business case. On the supply side, the aeroplane and car industries will probably generate enough supply of old airplanes and car parts. Getting all carbon fibre bicycle parts to the few facilities where carbon fibre is recycled, will require an effort from bike shops (as a collection point) and from distributors, who will need to set up a reverse logistics process to get the old carbon fibre to a recycling facility. Specialized has set up a Carbon Fiber Recycling Program in the United States to do this.[29]

The next question is, what sort of products can you make with the recycled material? The short fibres are less suitable for framesets, but for other parts that need to have a low weight but do not need to be quite as strong, they might be suitable. In the cycling industry, these parts could include soles of shoes, rims, sunglasses and helmets. A good design strategy would make it possible to reuse the fibres again and again.

A strong point of carbon fibre, which is often overlooked, is that it can be repaired by qualified companies. There are reports of framesets belonging to Continental teams that have been repaired and repainted multiple times during a single season because the teams did not have the budget to buy brand new framesets after each crash.

6. REGROW

In nature, nothing goes to waste. Everything, plant and animal, is part of a cycle where one species' remains become the nutrients for another species to grow. It is a very complex system and it has worked for millions of years, slowly adapting to changing circumstances. Birds fly at high speed without any noise, using minimal energy because of superb aerodynamics. Similarly, superb hydrodynamics allows fish to swim across oceans just to breed. Many trees have been around for longer than the first steam engines. A complex ecosystem of insects and even smaller animals breaks down dead plants to a natural fertiliser. Photosynthesis allows plants to harness the sun's energy. Nature has more examples of mind-blowing engineering than we humans will ever be able to come up with.

Nature is a great inspiration for circular product design as it is very good at solving numerous design challenges. 'Biomimicry' – the imitation of nature – is a way for designers to look at nature for inspiration. Early flying attempts were based on studies of bird wings, termite constructions keep the heat and the cold of the desert out, and spiderweb silk is one of the toughest materials on earth.[30] The self-healing properties of humans, animals and plants are a feature that would greatly benefit bicycle components. And trees and plants capture and store CO_2 from the air, one of the most effective and cheapest ways to reduce carbon dioxide levels in the earth's atmosphere.

Research into natural materials has not been as intensive as research into man-made technical materials. This is probably related to the complexity of natural systems, and a lack of financial incentives: nature is not interested in making money. Countries rich in resources as well as mining and fossil fuel companies are. They therefore invest in research into technical materials such as metals and oil-based ones like plastics to increase demand. However, there is a huge potential for the development of innovative, biobased materials.

Biobased materials offer a number of great opportunities for product design:

1. If grown and produced responsibly, they provide an infinite source of materials with the added benefit of regenerating the areas where they are produced.
2. Plants extract CO_2 from the atmosphere, storing it in the products that are made from them.
3. If produced with the material's end-of-life in mind, products can be returned to the biosphere to provide nutrients for new plants.
4. Some materials have equal or even better properties than the technical materials we currently use.

In the cycling industry, several water bottle manufacturers have started using alternatives to petroleum-based plastics, partly in response to negative press. Shown on TV screens all over the world, pro riders throw away their water bottles at the start of a climb to shed some weight. The colourful plastic bottle is sometimes picked up by a spectator along the road who will take it home as a souvenir. However, most bottles will disappear in roadside ditches, where they will take hundreds of years to decompose. Eventually the bottles end up in our food chain, as plastic particles are eaten by fish and other animals, and microplastics are consumed by plankton, all the way at the bottom of our food chain.

The new bottles are produced from biobased polyethylene made from sugarcane or potato starch. This material reduces the use of finite resources, requires less energy to make, and causes less pollution in the production process. These biobased materials are currently not strong enough to use for bottle caps, which are still made from traditional plastics. Even if the caps were also made from biobased materials, with 'biodegradable' properties, this still would not mean you could just throw away your water bottle and all will be good. The conditions in which current biodegradable plastics can be composted require high temperatures and a specific degree of humidity – a combination hardly found in nature. Instead, water bottles should be used as long as possible and then recycled in existing plastic recycling facilities to become a raw material for new plastics.

Three things have to happen to make biobased materials a serious alternative:

1. Create demand.
2. Initiate research & testing programs to develop more and cost-effective materials.
3. Devise a strategy how to make sure biobased materials actually become nutrients for nature, instead of a new stream of waste.

COMPENSATING CO2 EMISSIONS

Ideally, we would stop burning fossil fuels right now to prevent the climate from spinning out of control. Unfortunately, it will take decades before we have enough alternatives to oil, coal and natural gas to make this a reality. Apart from reducing your (or your company's) energy use and switching to renewable sources, one of the easiest things you can do now is to compensate your own emissions by planting trees. It will help conserving existing forests, create new ones, fight the expansion of deserts, and increase the required resources for biobased materials. The beauty is, there is no need to wait for politicians or technology to start doing this.

Research shows that there is a potential for more than 1 trillion additional trees to grow in areas that are both available and suitable, potentially removing 200 billion metric tons of carbon from the atmosphere. At the most effective locations, planting a tree costs as little as $0.30, so it would require $300 billion to make this immense project happen. With 7.8 billion people on the planet today, this is just $40 per person![31]

There are many ways to compensate your carbon emissions, and more are popping up every day. It is important to try and find a reliable party to do it for you if you do not have the opportunity to plant trees on your own property. More and more airlines, and even petrol stations, offer you the option to compensate for your emissions. Do it – the costs are minimal and the message to these companies that you care, is important. Another option is to calculate your yearly carbon emissions on the website of one of the many NGOs offering compensation schemes. Look out for local initiatives where you can contribute to the development of woods around your favourite riding area, or organisations certified to a Gold Standard, Plan Vivo or Verified Carbon Standard (VCS), preferably combined with the Climate, Community and Biodiversity Standard (CCB).

For companies and teams, the Greenhouse Gas (GHG) Protocol offers a global standard to calculate carbon emissions. This gives an insight into the main sources of the emissions and how to reduce these, as well as the amount that needs to be compensated for. Factoring future taxes on carbon emissions into your investment decisions helps to identify financial risks related to climate change. Starting with your own 'voluntary carbon tax' in the form of compensation is a good way to make your business conscious of this financial impact.

Even better than compensating last year's emissions is to compensate for all of your estimated emissions for the coming 10 years right now. If you do so, the trees are planted now and not later, allowing them to start taking carbon dioxide out of the atmosphere right now instead of in 10 years' time.

3.2 CIRCULAR BUSINESS MODELS

There are numerous opportunities in the cycling industry to save costs and increase revenues by using less material and using it more efficiently, you just need to open your eyes to spot them. The business models below fit different kinds of products and businesses; a combination works best in most cases.

1. REFUSE TO BUY

This is probably the most obvious and most widely applied way of saving costs for both consumers and manufacturers: not buying a product or service in the first place. The great benefit is that, if you do not buy something, it will not need to be made at all (an effect that will take some time for manufacturers to adapt to).

At a consumer level, refusing to buy new stuff every season to keep up with the latest fashion trend, for example refusing to buy a new cycling shirt and use the ones in your closet for another season until they are worn, will not only save money, but also materials.

Why this is circular: using less material for the production of new goods automatically reduces the environmental impact.

2. DURABILITY - MAKE IT LAST FOR EVER

A business model that is already quite common in the cycling industry is durability. If you make a product of superior quality, a higher selling price is justified. The beauty of this business model is that good parts and bad parts often have a very similar environmental impact in the pre-use phase, but because the former last way longer than the latter, their ecological impact over their entire lifetime is considerably lower.

Note that this business model doesn't work for every product. Cycling computer technology, for instance, is advancing so rapidly that there is no point in designing a housing for them that will last for decades.

Why this is circular: materials used in the product last longer than the average, reducing the impact per ride considerably.

3D printing of magnesium/titanium/carbon

Carbon with Dyneema® and (over)braiding

Personal fitted geometry, customised design & (human) validation

Carbon with Dyneema® and (over)braiding

Biobased glue

Integrated fibre sensing

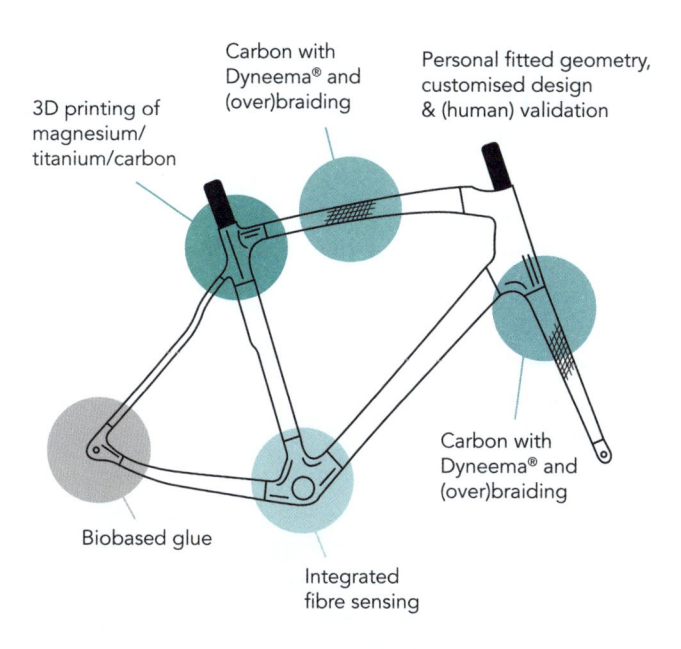

Figure 7: 100% Limburg Bike.

THE INDESTRUCTIBLE BIKE

More and more manufacturers make products that are so strong and durable that they are extending their warranty periods. Despite the serious impact carbon rims have to endure, producers like ENVE, Bontrager, ROVAL and Giant already make rims that are so strong that it is possible to offer customers limited lifetime warranties.

The Belgian Cycling Factory, owner of Ridley and Eddy Merckx cycles, joined a number of local production and knowledge partners in the '100% Limburg Bike' (Figure 7) project that is partly financed by the EU. The goal of the project is to create an 'indestructible' road bike that will not break in a crash thanks to the addition of unbreakable fibres to the carbon. The bike will be produced to order with a customised design and personalized geometry based on a bike fit, which will be realised with 3D-printed magnesium lugs combined with (over)braided tubes. Sensors will be integrated into the frame to monitor its condition, and biobased glues will replace traditional epoxies.[32]

3. THE FOUR RS: REPAIR, RESELL, REFURBISH AND REMANUFACTURE

Small businesses have been making money repairing, reselling, and restoring bicycles for as long as the bicycle has been around. Traditionally, it has always been a services business: independent professionals provide repairs and maintenance to bicycles of any brand directly to consumers. The bigger brands are hardly involved in these activities in the use and post-use phases of their own products. However, as these products become more reliable and consumers become more conscious of the environmental impact of their buying decisions, more customers will opt for reused, refurbished and remanufactured products.

For bigger brands, this offers the potential to set up a new line of 'pre-loved' bicycles alongside their completely new products. It would require a reverse supply chain through their existing dealer networks to gain access to their pre-used products, for example through a 'buy back guarantee'. Once back in the factory, thanks to the big brands' scale, efficiency and quality control systems, the bikes can be restored to a quality level almost equal to that of a new bike. Once the bikes are finished, the brands' trusted sales channels can be used in the same way as for new products.

Why this is circular: just like the design for durability business model, the 4 Rs retain value in the materials used for the product by increasing the number of rides per product.

4. SHARING - RENTING INSTEAD OF SELLING

Bike sharing schemes have been popping up all over the world in recent years. More and more people do not buy a bike, they only pay to use one. The idea of sharing a bike to increase its use makes a lot of sense: most bikes are only used for an average of a few hours a week. Increasing the use of a single bike by allowing several riders to share it, means fewer new bikes will be needed.

Serious bike sharing systems for road bikes give you the freedom to go out for a ride any time you feel like it, with bikes parked very close to you. However, this business model seems unfit for a general

TRAINING CAMP BIKE RENTAL

Providing access to a bicycle through a rental system can be very lucrative. It is very popular in tourist destinations that offer great riding far from home. Places like Mallorca and the Canary Islands all have a thriving bike rental industry with bikes in all shapes and sizes. Riders are willing to pay a fairly large amount per ride, because it saves them the hassle of taking their own bike on a plane. They get a bike of really good quality for the duration of their holiday.

It is no coincidence that rental companies select bikes that fit the 'design for durability' or the 'resell' category, minimizing the required maintenance of rental bikes, and resulting in the highest possible resell value at the end of the season, when the entire fleet is often replaced.

rollout, as the personal preferences of road cyclists are very diverse. A single one-size-fits-all road bike that works for everyone is almost impossible to design. Apart from its size, the 'identity' of a road bike is often an important part of why a rider rides it – the bike you ride is often part of your status.

There are more interesting opportunities when you start looking at parts, tools and apparel. Most riders have many expensive parts and clothes in their sheds and closets. Stuff that they buy for a specific event or maintenance job, and that is hardly ever used. Take wheelsets for example. Many riders own a couple of wheelsets, one for flat roads and another one for when they hit the mountains. If a rider could rent a high-end climbing wheelset for a couple of trips a year, this would not only be economical – because no investment is required – but also save space in their shed. Additionally, second wheelsets often become obsolete way before they are worn, e.g. when someone gets a new bike and the wheelsets are no longer compatible because the new bike has disc brakes.

A second opportunity is renting a special, and often expensive, tool for a weekend to, for example, replace your bottom bracket. You would select a tool and an instruction video on a website, receive the tool by mail, and return it in the original box once you are done with it.

Expensive clothes and other gear made specifically for certain conditions is our final example. Consider, for example, that waterproof and super-breathable rain jacket that you will really only need for that special spring trip to the mountains with poor weather forecasts. Why spend several hundreds of euros on a jacket you end up using for only one afternoon? Renting it would have saved you money and made sure the materials were shared by more riders.

Why this is circular: in specific situations, providing access to a product for only a limited period of time makes economic sense for a user because renting is cheaper or saves space. The rental process needs to be easy and reliable, but if it is, it saves a lot of materials that are now underused.

5. PAY FOR PERFORMANCE

It simply works – and it only becomes an issue when it does not. Most users are only interested in the performance of a product, not so much in what it is or what it looks like. It's like getting from A to B dry and safe by taxi: you tell the driver where to pick you up and drop you off, and you pay for the service, provided it is good enough. If the taxi breaks down, it is not your problem, you simply find another to continue your journey.

Taking this idea one step further, you would want the original manufacturer of a product to be responsible for its performance over its entire lifetime, i.e. during use and post-use. Instead of buying a new product every time a product is worn or broken and needs to be replaced, a user pays a fixed monthly fee for the performance the product delivers. The manufacturer gets a constant flow of income, which provides the ultimate incentive to make sure a product lasts for as long as possible, with as little maintenance as possible.

It also makes the manufacturer responsible for the product at the end of its lifecycle, creating opportunities for reuse, refurbishment, remanufacturing and responsible parts recycling. Since the user pays for a product that works, manufacturers have the option to mix new and used parts, as long as the product does what the customer pays for.

Interaction with customers in a performance model is completely different from the traditional sales model. A constant feed of information (as well as money) flows back to the manufacturer: the use of the product, the way it functions, and how future products could be improved to increase durability and reduce maintenance requirements. It will allow manufacturers to save costs by keeping materials in use for as long as possible, maintaining value on the highest possible level, and at the same time reducing the use of new resources and of waste.

Why this is circular: pay for performance makes manufacturers responsible for the value of a product over its entire lifecycle. It provides the best incentive to design for durability, ease of maintenance and repair. It offers the option to refurbish a product for another lifecycle and, because the product remains the property of the manufacturer, all the parts return to the factory for remanufacturing or recycling.

6. HYBRID MODELS

Hybrid models combine a product that lasts a long time with parts with a limited lifetime, the so-called consumables. Instead of disposing the entire product, only a part of it needs to be replaced, limiting the economic and environmental value loss. The accompanying business proposition is often the sale of the long-lasting product with a first set of consumables, followed by returning sales of the consumables over the lifetime of the product. Coffee machines that take specific capsules, printers using replaceable cartridges, and vacuum cleaners that come with replaceable bags, are examples of hybrid models that are frequently found around the house or in the office. Each product consists of a clear base product and a set of consumables.

Clipless pedals are possibly the best-known example of a hybrid model in the cycling industry: pedals last far longer than cleats mounted under shoes, which get all the abuse and wear out quickly. Instead of buying a new set of pedals and cleats, you only buy new cleats. Brake pads and rear derailleur pulley wheels are other examples of parts that wear faster than other parts of the brake or derailleur. Both have been designed to be replaced easily at a fraction of the cost of the total product. Note that all these examples do not have return options or clear disposal instructions, meaning there is room for improvement.

The trick is to get the price level of both the base product and the consumables right. Make the base product too expensive and consumers will look for an alternative. Make the consumables too expensive, to compensate for a discount on the base product, and other suppliers will start to sell cheaper, compatible consumables. A close relationship with customers is important to prevent them from choosing alternative consumables. Hybrid models should come with consumables that have a low environmental impact, because of the short lifecycle of these materials.

Why this is circular: the hybrid model creates a 'lock-in' for the user. Once the first investment in the long-lasting product has been made, a constant flow of sales of the consumables is guaranteed to the supplier over the lifetime of the long-lasting product. The longer the life span of the base product, the longer the supplier can sell the consumables. This creates an incentive to develop a very long-lasting base product and consumables that require little material to save on recurring costs. Provided there is also a return incentive for the worn or empty consumables to make sure these materials are reused or recycled, the hybrid model keeps materials in use for a long time.

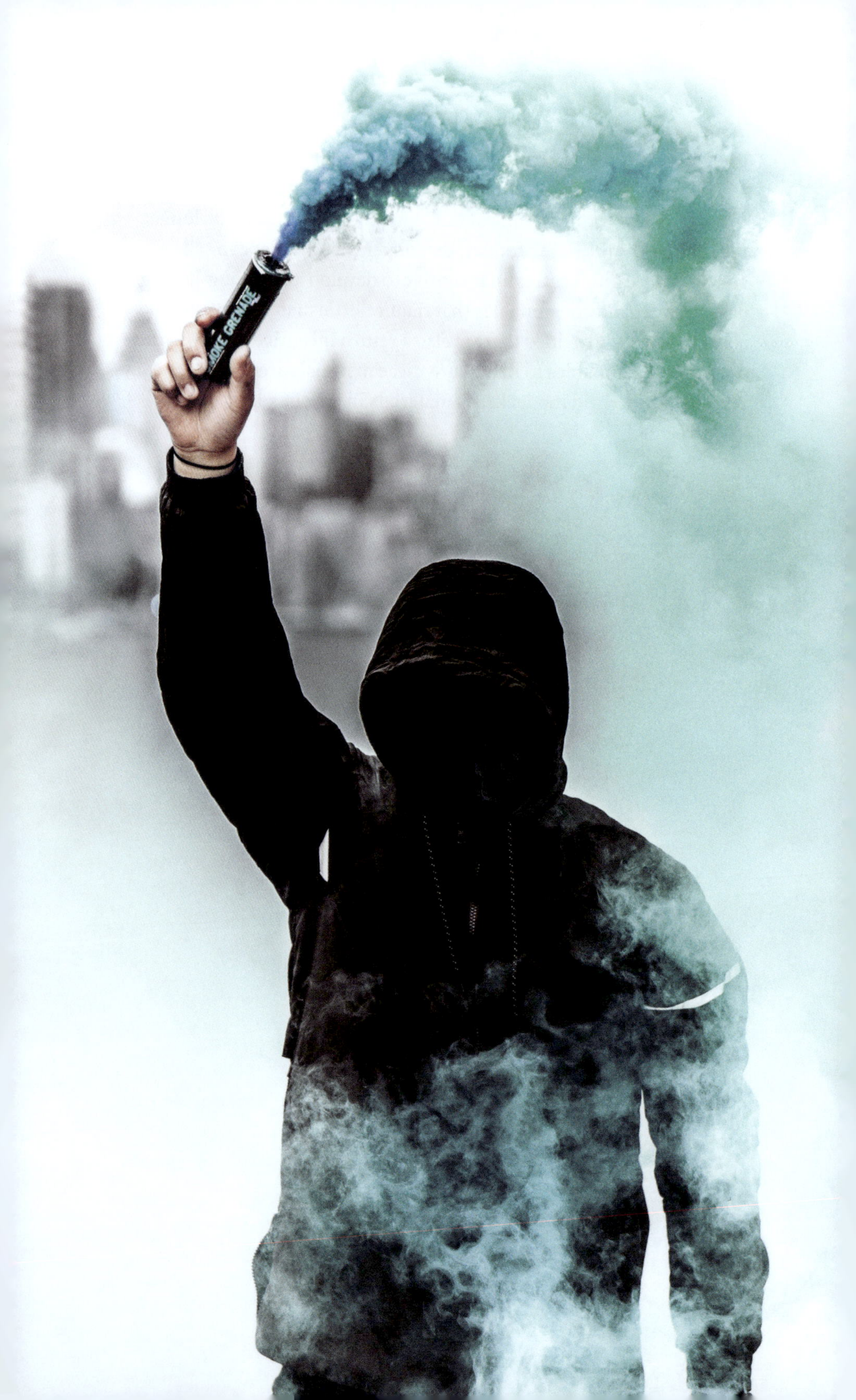
OKE GRENADE

STAGE 4
ACTION PLAN FOR A CIRCULAR CYCLING INDUSTRY

We are entering dangerous territory now. We do not have all the knowledge and ideas, nor the money and power, to successfully make the transition to a circular cycling economy on our own. We do not want to blame anyone for the situation we are in today, because we are all part of the same system. Nor do we want to dismiss the creativity of the cycling industry. With our ideas, we would like to contribute to an industry that will be able to make a fair profit, without causing waste and pollution. The business models we introduce are based on the assumption that customers will not change the amount of money they will spend on a bike, but that they will spend their money differently.

We feel that it is worthwhile to share our experience to provide inspiration and kick-start discussions within the industry. To do this, we present our 'draft action plan' for a circular cycling industry as first input for an industry-wide, shared vision, and some ideas about the contribution each stakeholder could make.

OUR LONG-TERM VISION

Bicycles, equipment and clothing are made from renewable plant-based materials or reused and recycled parts. Materials wearing from tyres or brake pads are biodegradable and lubricants provide valuable nutrients for the environment.

Materials are in use longer because of longer-lasting designs, maintenance is easier, and wear indicators tell users when parts are worn and up for replacement. Repairs can be done efficiently and are part of manufacturers' business models. Used bikes, bike parts, equipment and clothing are returned to their manufacturer, so they can reuse parts and materials to make new products. Or, alternatively, materials are biodegradable, so they can become part of the circle of life again.

Factories, offices and transportation throughout the industry run on renewable sources of energy. Packaging materials go back and forth between manufacturers and users to protect new parts on their way to users and used parts on their way back to the manufacturer. Pollution and waste are a thing of the past, and nature gets a chance to recover, so it can provide us with even nicer places to ride.

It would be a win-win-win situation. Customers win because they are able to ride more reliable products that require less maintenance. The industry wins because it can create equal or more value with less material. Finally, the planet wins because there will no longer be any waste or pollution, and nature will get a chance to regenerate.

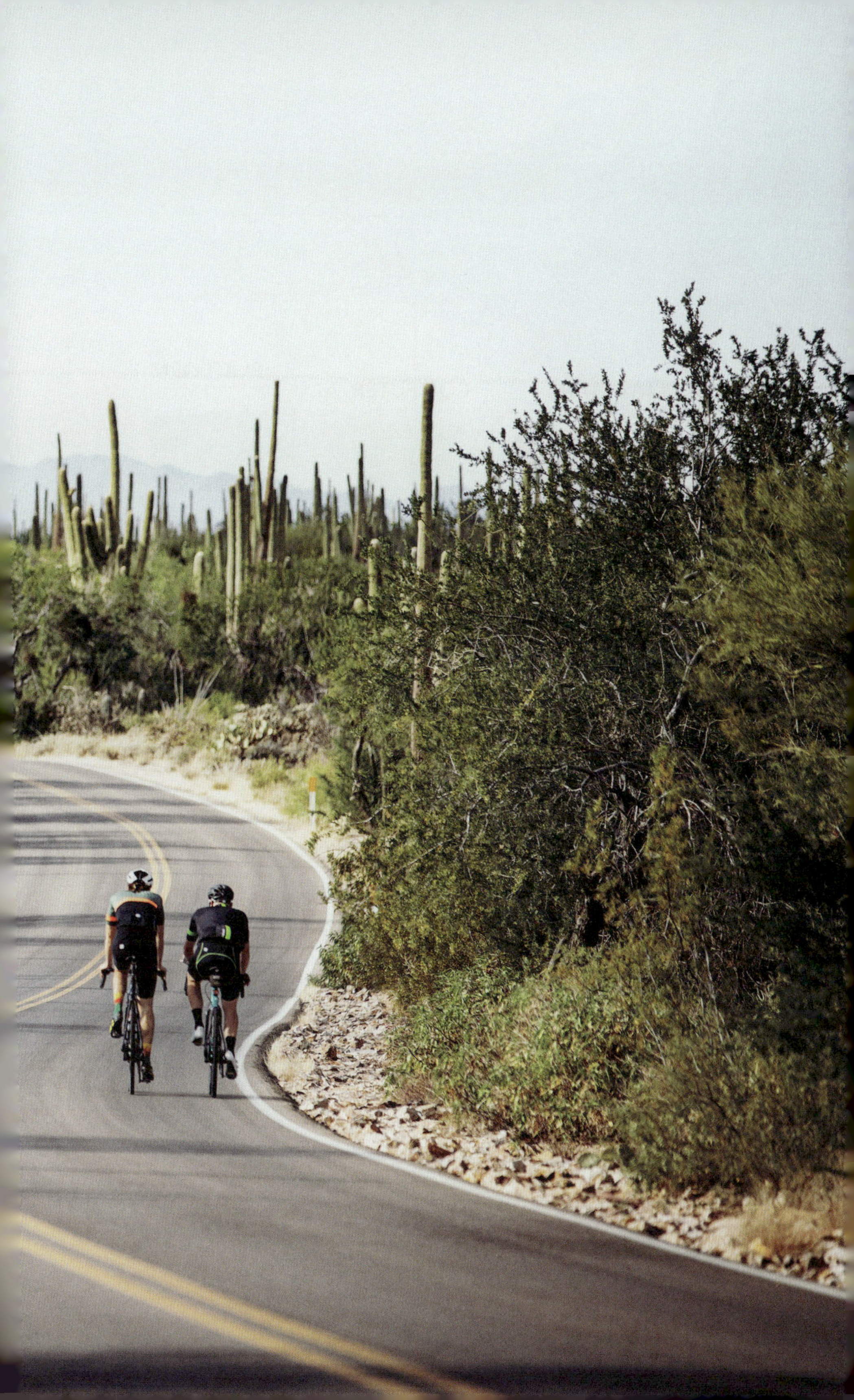

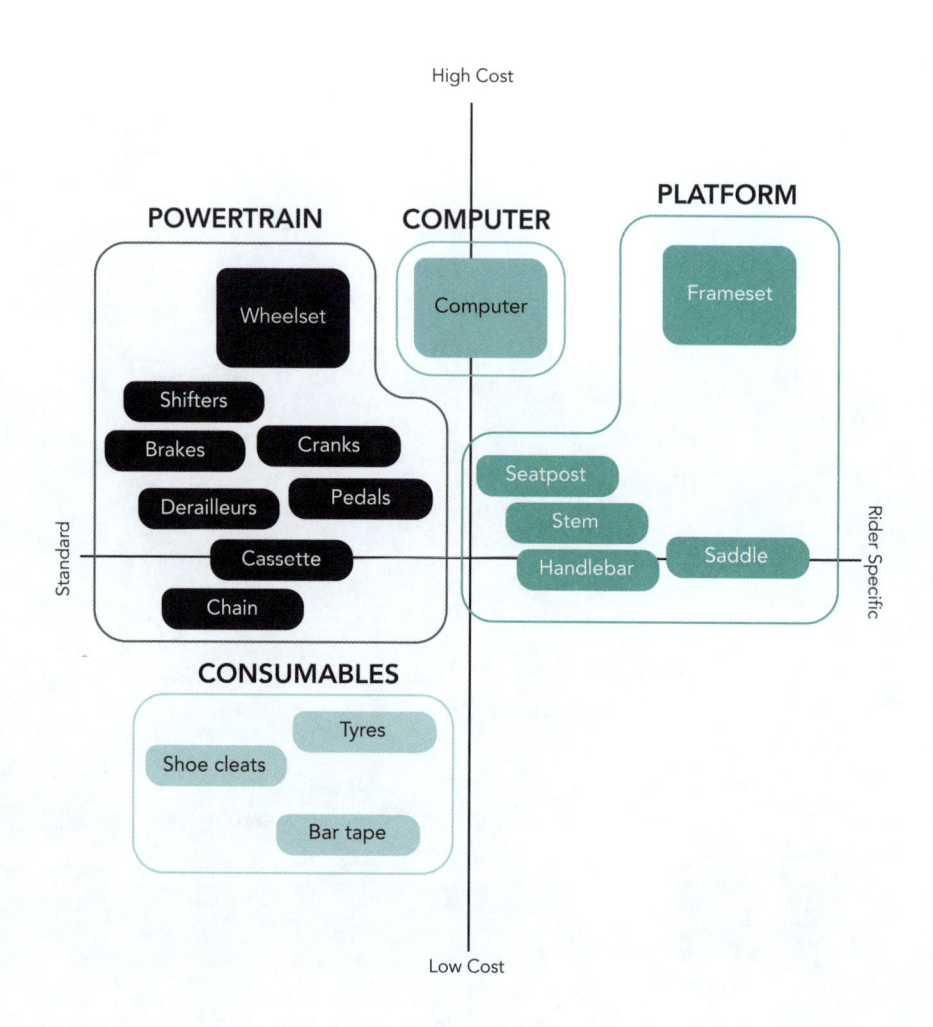

Figure 8: Classification of components for circular business models and design strategies. On the vertical axis the price of a component, on the horizontal axis the difference between standard and rider-specific properties.

4.1 THE BUSINESS MODEL OF THE CIRCULAR BIKE

As you can imagine, keeping materials in use longer in the way we described, will have an enormous impact on existing business models. The current linear models are based on the principle of generating as much profit as possible from selling as many products as possible. This means that the transition to a circular economy cannot be realised by designers alone – it requires the development of new ways of doing business with both suppliers and customers.

The most suitable circular design strategy and business model for a particular bike component depend very much on the function, the cost and the lifetime of the component. Some components are rider specific, like the size and looks of a frame, others like a chain or brakes are not. We have categorised the main bike components in Figure 8 to show how rider specific each item is compared to a standard component, and what the price range is for each item.

Based on this classification, we identified four product groups with their own specific business models and design strategies: Platform, Powertrain, Computer and Consumables.

The Platform consists of a frame and fork, headset, stem and handlebar, and seatpost and saddle. It is the first thing a customer selects when searching for a new bike. A Platform is rider specific, long lasting and costly. The traditional groupset containing, for example, shifters, derailleurs and brakes, as well as the wheels and pedals, are grouped in the Powertrain. The parts are fairly standard, prone to wear, and relatively costly when they need to be replaced. Easy to replace and relatively low-cost parts like the tyres, shoe cleats and bar tape together form the group of Consumables. The final product group is the Computer.

Each product group has its own business model and design strategy, which we will explain in more detail later on. Compared to the existing linear business model, the customer no longer pays for the entire bike at once when he or she orders the bike at a shop. Instead, the complete bike will have a circular hybrid business model, where customers buy only the Platform and Consumables, and pay a monthly fee for the use of the Powertrain and the Computer.

In Figure 9, we have plotted the cumulative cost of a €3,000 traditional bike and our proposed circular bike, based on a lifetime of six years. In the linear model, the customer pays €3,000 at the moment the bike is purchased. Over its lifetime, the bike will have to be maintained and repaired, and worn parts will be replaced several times, often at irregular intervals. At the end of the lifetime, the bike can be sold on the second hand market. In the circular hybrid model, the initial cost of a bike is lower than in the linear business model. The customer will buy the Platform and the Consumables, but the Powertrain and Computer will remain property of the manufacturers, in return for a regular fee based on the use of the product, as part of a long-term contract.

Over the lifetime of the product, the amount of money spent by the customer in the circular model will be comparable to what he or she would spend in the linear model. The main differences are that a rider has a lower initial investment and then pays a fixed monthly fee for the risk-free use of the Powertrain (all maintenance costs are covered in the contract) and Computer, and will have a guaranteed buy-back price for the Platform at the end of the lifetime.

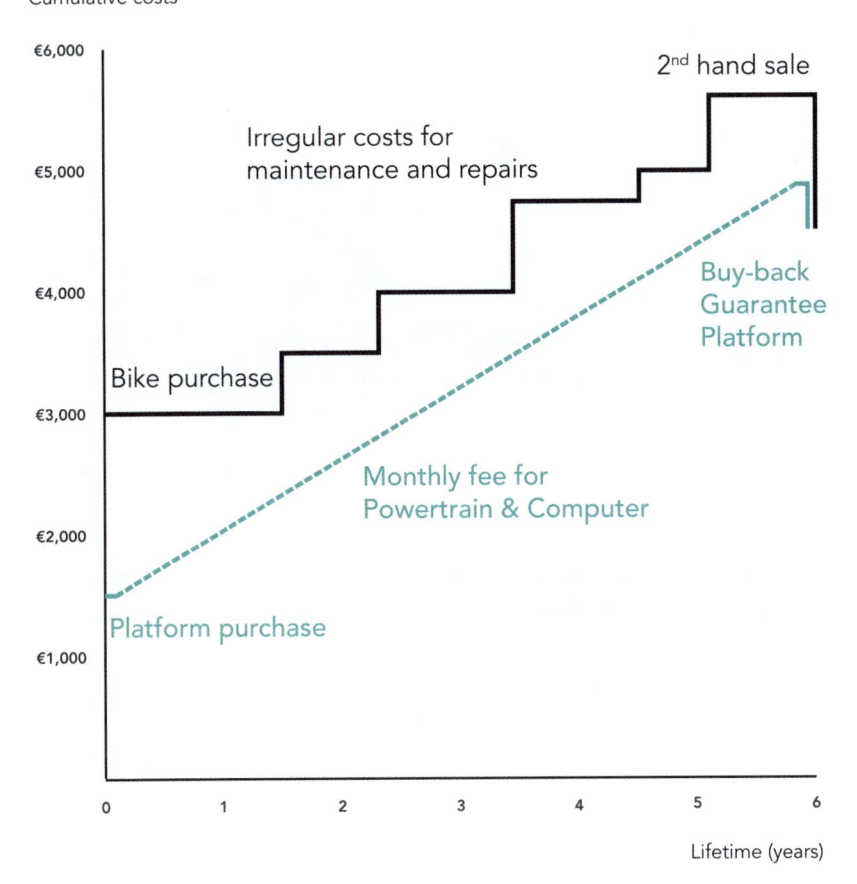

Figure 9: Cumulative costs of a road bike over the lifetime (continuous line existing business model, dotted line proposed circular business model).

I. PLATFORM

The 'heart & soul' of the road bike is the Platform (Figure 10). It is what distinguishes a rider's bike from the others, a billboard reflecting the ego of the owner. To own a bike really is to own a Platform – this is what will last, while Powertrain components come and go as they wear or break. Consisting of the frame and fork, headset, stem and handlebar, and seatpost and saddle, the parts making up the Platform determine the fit and an important part of the bike's ride quality and riding style.

Figure 10: The Platform consists of a frame and fork, headset, stem and handlebar, and seatpost and saddle.

Current knowledge and materials already make it possible to design and manufacture Platforms that will last forever, because they are strong and hardly wear. This makes it possible for Platforms to be used and reused by several riders over a long period of time, which is the basic idea behind the circular business model of the Platform. It implies that all Platforms will become top-end Platforms. As we experienced building our UpCycles, even a 10-year-old top-end frameset can provide a much better ride than a new low- to mid-level frameset. Ideally everyone would be able to ride a top-end Platform, and this circular business model could make this happen.

Platform manufacturers would offer three product categories at different price levels:

1. **Latest technology, custom paint:** based on the latest insights in bicycle design, these Platforms will be made from recycled and/or renewable materials, and come with a custom colour scheme;

2. **Pre-used, custom paint:** consisting of pre-used components that have been checked, repaired when required, potentially updated with new features, and painted in a custom colour scheme;

3. **Pre-used, original paint:** consisting of pre-used components that have been checked and are ready for another lifecycle without the need for repairs or repainting, even though the paint might have some scratches from a previous lifecycle.

Figure 11 shows the flow of materials in a Value Hill. Manufacturers' design capacity and production will be focused on the development of new Platforms with the latest technology, similar to the current top-end models. Instead of designing a whole line-up of bikes from high-end all the way down to entry level, the design effort will be limited to

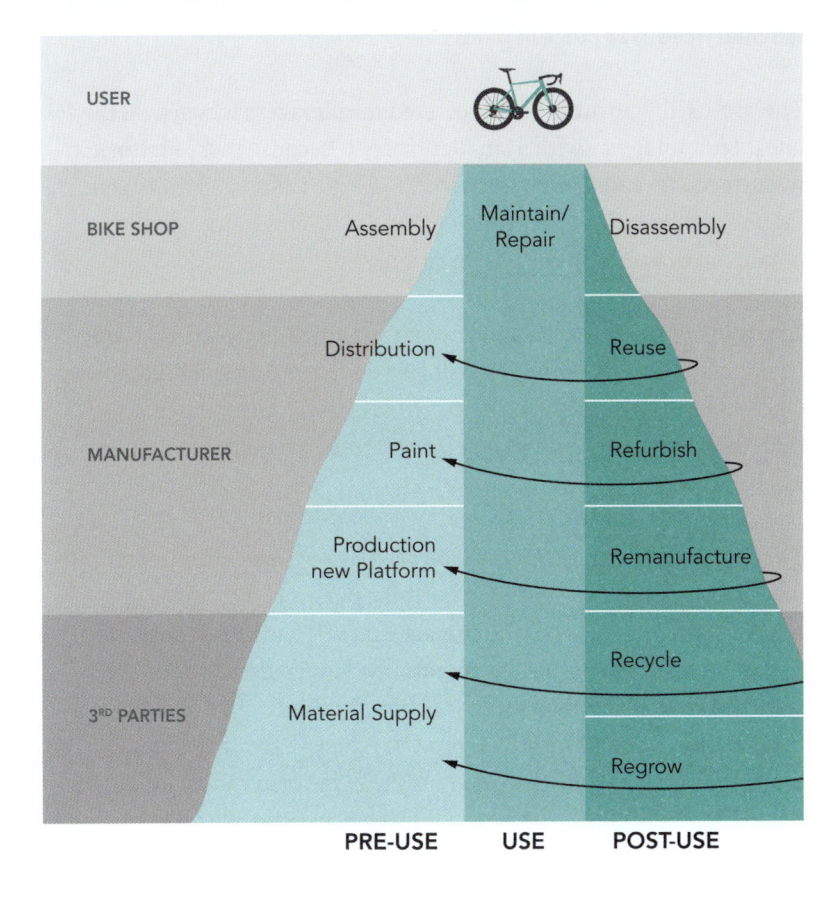

Figure 11: Platform Value Hill.

a number of options for various riding styles, like climbing, aero and general purpose. Every new product leaving the factory will be of the highest quality, will have the best riding characteristics, and will be made from remanufactured, recycled or biobased materials.

The buy-back guarantee provided by the Platform manufacturers will develop a flow of used products back to the manufacturers. This gives them the opportunity to reuse or refurbish Platforms. Each returned Platform will go through a thorough and highly automated quality control procedure at the manufacturer's facility. Depending on the state of the returned Platform, it can be forwarded to the inventory immediately, and be labelled 'ready for reuse', or, if it is damaged or severely scratched, it will be submitted to a refurbishment process for small repairs and a new paintjob, and be labelled 'ready for custom paint'.

The pre-used Platforms will be offered to riders less focused on the latest technology or with a smaller budget. Because these Platforms were once top-end models, customers will be riding a Platform with far better properties than the lower- and mid-end framesets they would have bought in the old linear line-up.

Assembly of the Platform and Powertrain into a complete bike will be done by a bike shop, which will also do the required maintenance. Once a manufacturer buys a Platform back from a customer, a bike shop will disassemble the bike and return the Platform to the manufacturer.

The integrity of a Platform is monitored by sensors communicating with a Computer, creating a log in a bike passport (see boxed text *the Bike Passport*) with use and impact data. This makes it possible to warn a rider when a part is potentially damaged and should be returned to the bike shop for inspection. This information will also benefit the reuse and refurbish process when a Platform is returned to a manufacturer. If, for whatever reason, a Platform can no longer be reused or refurbished, the manufacturer is responsible for recycling or returning the materials to the biosphere to become nutrients again.

Business model
The Platform concept comes with a 'durability' business model, where manufacturers focus on making long-lasting products. Setting up a new business line focused on reuse and refurbishment, will allow manufacturers to increase their own revenues, and at the same time reduce the size of their supply chains and associated costs. Additionally, having to deal with fewer suppliers will decrease the risk of supply chain shocks.

Each Platform becomes part of a custom bike. This means that there is no longer a need for the yearly rat race to launch an entire new line-up, which needs to be in the stores before the season starts and be sold before it ends, often in an end-of-season sale. The total value (and risk) of the inventory is further reduced because the Powertrains will no longer be part of the package sold to bike shops.

Like the manufacturers, the bike shops will have lower inventory costs, because there will be no need to hold stock of assembled bikes before they are sold. Bike shops will make money where they can add real value: advising customers, carrying out bike fits, assembling bikes, doing repairs, and disassembling the bike to return the Platforms and Powertrains to their respective manufacturers. Shops will make a margin on a Platform just like they do on the sales of a complete bicycle today. The Platform design will be such that both assembly and disassembly can be done quickly and reliably, and IT systems will support salespeople and mechanics, leading to a reduction of overhead costs.

Pricing
The way Platforms are sold is much like how bicycles are currently sold, via a bike shop or directly to a consumer via the manufacturer's website. The custom paint models will be painted in a scheme chosen by the customer, applied overnight and delivered to the bike shop within a matter of days for assembly. The pre-used models without custom paint potentially have a few spots where the paint is scratched or damaged from a previous lifecycle.

The table below gives an idea of the price levels of the different Platforms and the expected buy-back:

	PRICE	BUY-BACK
Latest technology, custom paint	€€€€ €€€€	€€€
Pre-used, custom paint	€€€ €€€	€€
Pre-used, original paint	€€ €€	€€

The buy-back guarantee is offered to the user to make sure the Platform is returned to the manufacturer. Bikes that have been damaged through improper use or accidents can be bought back at a discount, to make sure the materials are still returned to the manufacturer to recycle and/or regrow them.

THE BIKE PASSPORT

The lack of accessible information was one of the main reasons why our Circular Cycling business proved so time consuming and labour intensive. We had to spend too much time searching for information about compatibility and installation manuals, as well as finding spare parts to achieve our target for the number of hours to create an UpCycle. We searched manufacturers' online portals, found manuals on obscure websites, and got sucked into 15-year-old forum discussions with all sorts of tricks to make stuff work together. We were often unable to find what we were looking for and thus forced to experiment. With the current cost of labour in a country like the Netherlands, it is hard to make the business case work.

What happens when researching parts is even worse when it comes to materials. No consumer or bike shop is able to identify which kinds of materials are used to produce parts, which means that proper disposal in the correct recycling stream is almost impossible.

Information is a key element of a circular economy. To be able to efficiently repair, refurbish and recycle a bicycle, reliable information needs to be easily available. Somewhere in the supply chain, all this information is available, since it is required to manufacture parts and assemble the bikes efficiently, but in the current situation it never reaches the bike shop or the consumer.

One of the first things we did when we set up Circular Cycling was the development of a Bike Passport. We made unique QR code stickers that we placed on our UpCycles. Scanning the code using a mobile phone gives access to the Bike Passport, an online location containing all the information about the bike we were able to find, ranging from its history to a detailed description of the parts and the estimated remaining lifetime when we sold the UpCycles. We used standardised descriptions of component fittings wherever possible, and stayed clear of marketing terms.

The concept of a bike passport makes it easy for users, as well as bike shops, to figure out which replacement parts are required, saving them the hassle we had to go through. Our vision of the bike passport goes way beyond the basic version that we made for our UpCycles. It would not only contain the straightforward information that we included, but also invoices, bike fit measurements, torque settings, installation manuals, production locations and dates, and warranty information. You should be able to go a level deeper and 'zoom in' on components for technical manuals and replacement part numbers. At the end of a part's technical lifetime, the bike passport will help you to figure out out the best way to recycle the materials with up-to-date local information about recycling options.

The passport is also the location where all the data from future sensors will be collected, ranging from standard data such as total distance, weather conditions and power, to new sources such as stresses in frame materials and the wear of bearings. The data in the passport can be used to send maintenance alerts to the users and provide the manufacturers with performance data on a scale never seen before. If Powertrain manufacturers provide their products in a 'pay-for-performance' business model, they can use the data to predict required maintenance and repairs, as well as track the use of parts that are returned to their factories after a contract ends. When replacing parts, scanning the barcode on the replacement part automatically updates the passport, including the date of the replacement. A bike passport would be a 'digital twin' of the bike itself.

In 2019, Cannondale introduced a bike passport, which is a big improvement over our basic version. It is available as an app on your phone, which pairs to a front wheel GPS sensor, logging the use of the bike. After scanning the frame number of your Cannondale, you can enter your bike fit information, check manuals and maintenance videos, and record your maintenance history. [33] Parallel to the app, Cannondale developed an Augmented Reality feature that appears after scanning a code on the bike using a tablet or smartphone. The 3D augmented version of the bike on the screen gives access to exploded views, maintenance manuals and lists of replacement parts.[34] Another example is the latest version of the SRAM AXS Web that already combines data gathered by the wireless SRAM electronic groupset about gear use and remaining battery life, with GPS-recorded speed and distance, as well as power meter data and tyre pressure measured by another set of sensors.[35]

The business opportunities of a bike passport are endless. Cheaper Internet of Things (IoT) sensors will make it possible to gather more and more information. Apps in the car industry, food delivery and for car sharing already provide some very good examples of similar digital transformations.

A bike passport could enable riders to spend more time riding their bikes instead of keeping their bikes in shape. The real boon for the industry is that a bike passport will become a goldmine of data. Initiatives such as the DST Foundation, where various industry partners work together on a standard for product data and an online database containing up-to-date information with API interfaces for shops and suppliers, are paving the way.[36] A bike passport will need to be governed properly to prevent it from becoming a 'data monopoly' where a single operator becomes too dominant. On the other hand, a proliferation of various bike passport standards is to be avoided as well.

Getting the balance right between short-term investments, open data and privacy is key for long-term success. Privacy especially is something to organise properly right from the start, to prevent data breaches and privacy concerns. We believe that privacy is another 'inconvenient truth' that in the next decade or so will result in even stricter rules and regulations than the recently introduced GDPR, as well as a consumer backlash if companies are unable to create a safe environment. Making sure the bike passport is future proof requires smart databases and strict rules on the kind of data that can be shared with whom.

The opportunity is huge, if the development is done properly.

2. POWERTRAIN

The parts of a bicycle that are most vulnerable to wear and damage are the parts of the groupset, wheelset and pedals. They transfer the power from the rider to the road. At the same time, these parts have to deal with rough road surfaces, shifting under heavy load and the weather, causing them to wear and fail over time.

Figure 12: The Powertrain consists of a groupset, wheelset and pedals.

Compared to the Platform, the Powertrain (Figure 12) has only a limited effect on the looks of a bike and personalised fit. The gear ratios currently on the market make it possible to ride on any terrain, without the need to change cassettes or chainrings. Disc brake wheels with carbon rims are such a common feature that they too no longer make a real difference in what a bike looks like.

All a rider really wants is to use the Powertrain without having to bother about maintenance, parts failing at the most unwelcome moments, and unexpected bills associated with these breakdowns. It is a perfect product group for a 'pay-for-performance' business model. This means that, unlike the Platform, the users will not actually own the parts and wheels, but will pay a monthly fee for the use of the products instead, which also covers maintenance and repairs.

During the process of choosing a bike in a bike shop, the rider picks a Powertrain bundle, in accordance with his or her expected annual kilometres, riding style and budget. The Powertrain bundle consists of a groupset, a wheelset and a set of pedals, which can be selected from a number of options with different monthly prices.

The Value Hill for the Powertrain can be found in Figure 13. For manufacturers, the most important thing is to offer different Powertrain options with enough reliability to provide the customer with the performance he or she is paying for. It means that manufacturers will have three main departments: collection & quality control of used parts, production of new parts, and a place where pre-used and new parts are combined into complete Powertrains ready for shipment to bike shops.

New technology will be introduced through parts that will replace parts that cannot be reused or refurbished. Just like with new Platforms, these new parts will be made from remanufactured, recycled and biobased materials.

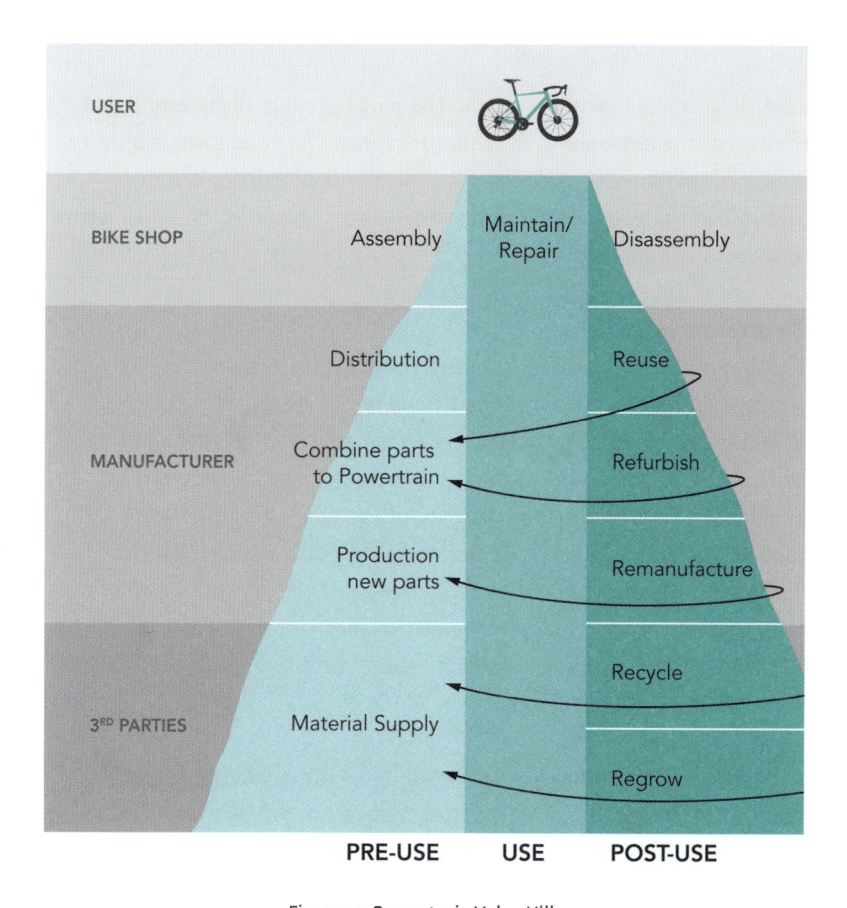

Figure 13: Powertrain Value Hill.

Bike shops are vital for Powertrain manufacturers. Shortly after the order for a new bike is completed, a manufacturer will send the Powertrain to a bike shop for assembly on the user's Platform. Each time a user visits the shop for maintenance or a repair job, the shop will be paid by the Powertrain manufacturer. This means that manufacturers will do everything to make sure that users can extend the lifetime of the parts themselves. The sensors in the parts and bike passport app on the user's phone will tell the user that he or she can 'save x number of watts on their next ride' by cleaning and lubricating his or her drivetrain, and show an instruction video on how to do this. Environmentally friendly chain lubricant is sent directly to the home address of the user when her or she runs out. If maintenance or repair by a bike shop is required, information from the bike passport will enable the manufacturer's computer system to automatically supply a selected bike shop with appropriate spare parts and instructions.

Bike shops return worn parts and the parts that are disassembled at the end of the contract to the manufacturer. The collection & quality control department of the manufacturer will determine the next lifecycle of the product through reuse, refurbishment, remanufacture or recycle and regrow.

Business model
The 'pay-for-performance' business model is different from anything we have seen in the cycling industry so far. In return for a fixed monthly fee, the user gets a system that transfers the rider's power from the Platform to the road. The user can expect the Powertrain to work without having to worry about the cost of maintenance and replacement parts during the period of the contract, as these are included in the monthly fee. Part of the contract is an obligation to keep the Powertrain clean, lubricate the chain, and have the bike checked on a regular basis by a bike shop or mobile mechanic visiting your home.

The Powertrain manufacturers organise all the financial transactions: the user is charged automatically every month; the bike shop gets paid for assembly, regular maintenance, repairs and disassembly. Tools and components required by a bike shop for work on a Powertrain, and the recovery of components at the end of the contract, are all provided by the manufacturer.

The manufacturers benefit from a reliable and constant cash flow. As they remain owner of the components, they become responsible for the entire lifecycle. To improve margins, manufacturers will have to make sure users go to a bike shop as little as possible for reasons other than ordering a new bike or planned maintenance: every time a rider needs to visit a bike shop because something does not work, a manufacturer loses money.

This gives manufacturers a real incentive to design parts that last as long as possible, and to provide the best performance at the lowest possible cost. This can be achieved through a combination of new designs with improved durability and more efficient assembly and repair procedures.

Because users pay for performance and not for a product, manufacturers can mix new, used and refurbished parts, all with a logged usage history and a reliable remaining lifetime estimated on the basis of data collected in the bike passport. As components are returned to the manufacturers at the end of the contract, components that cannot be reused or refurbished can be fed back into the production line for remanufacturing, collected for recycling or returned to the biosphere.

Additional revenues can be realised, for example, when a customer wants to use a lightweight wheelset for a short time or upgrade to a lighter groupset. For a small one-off fee plus a higher monthly charge, the bike is upgraded and the parts coming off the bike are returned to the manufacturers to be used on another bike.

Pricing
The pricing of the pay-for-performance business model needs to be diversified depending on the selected options and the annual use. As distribution costs and labour rates vary depending on the geographical location, the price of the bundle may also differ per region.

Similar to choosing a mobile phone bundle, a user makes an estimate of the annual use and selects options for the Powertrain that fit his or her budget. If a user exceeds the estimated yearly kilometres, an additional per-kilometre fee will be charged automatically, based on the information collected in the bike passport.

The table below gives an indication of what a bundle could look like. Just like with mobile phones, additional items, such as a power meter, can be added to the bundle.

	0 – 3,000 KM PER YEAR	3,001 – 7,500 KM PER YEAR	7,500+ KM PER YEAR
Powertrain Basic	€€€/KM	€€/KM	€/KM
Powertrain Medium	€€€€/KM	€€€/KM	€€/KM
Powertrain Pro	€€€€€/KM	€€€€/KM	€€€/KM

The transparency of the bundles will give consumers better insight into the total expected costs for the entire duration of the contract for various suppliers, boosting competition between them to provide long-lasting products at the lowest cost for consumers.

3. COMPUTER

The Computer will be the bike's data hub. Current computers collect a rider's health & performance data, such as speed, cadence, heart rate and power output. In a circular business model, the bike's health will be monitored too, using an array of sensors and wireless chips sending information to the Computer and then to the online bike passport. Riders will regularly get a report of the status of the bike, for example when a part approaches the end of its lifetime or when batteries require a charge. In bike races, the computer could even send information about a rider with a broken bike to the team car, so mechanics know early on what the problem is.

A rider will choose a Computer when picking the Platform and the Powertrain at the bike shop. Standard protocols for communication between the sensors in the bike parts and the Computer, and for communication between the Computer and the online bike passport, ensure that any Computer can be used, independent of the combination of the Platform and Powertrain chosen by the customer. As mentioned before, data protection and privacy need to be at the core of the data flows for the bike passport to work in the long term.

Business model
Just like a Powertrain, a user wants a Computer to work. A pay-for-performance model would work best to make sure Computers will work for as long as possible. It would make the manufacturers responsible for, for example, the performance of the Computer after software updates, battery performance, and the interfaces between the Computer and the various sensors on the bike. In return, the user will pay a monthly fee for the use of the product.

4. CONSUMABLES

The Consumables are the parts that wear relatively quickly and can easily be replaced by the user, or at a bike shop if the user does not want to get dirty hands. Consumables include shoe cleats, tyres, brake pads and handlebar tape. Shoe cleats, tyres and brake pads wear because they come into contact with the environment – cleats when walking, tyres when riding, and brake pads when braking. The material that wears from the parts ends up as waste, directly polluting nature, the worn-out remains are often thrown into the waste bin.

The challenge for the manufacturers of these parts is to make them from renewable, non-toxic materials that become nutrients for plants when they wear. Worn-out products should come with a clear instruction of how users can make sure they do not end up in a landfill or incinerator. If the worn parts cannot effectively be disposed in local recycling systems, local bike shops could function as a collection point to create enough volume for economic recycling. Because the Consumables are registered in the bike passport too, ordering replacements can be done directly through the bike passport app at a brick-and-mortar or online shop that offers the best deal.

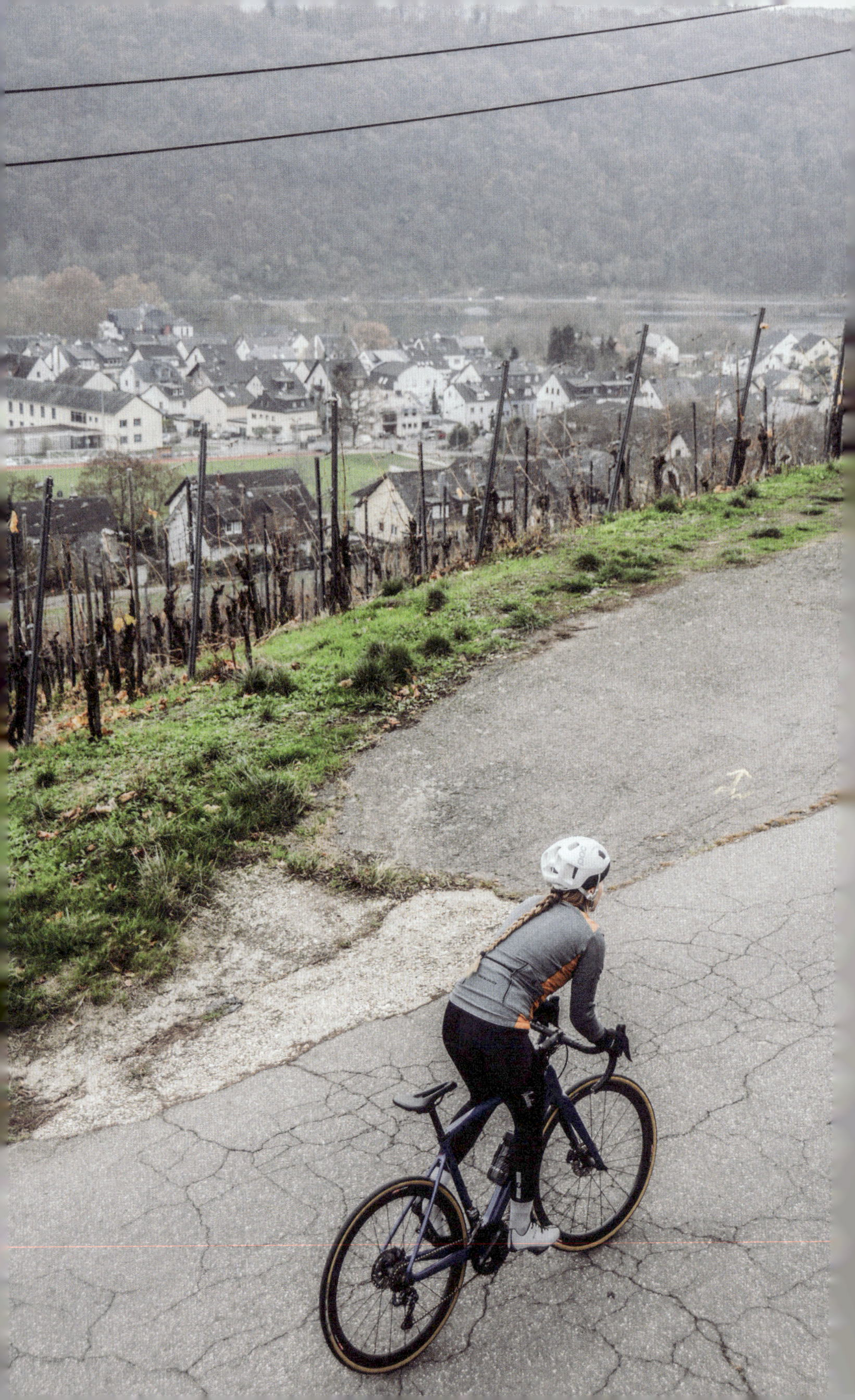

Business model
The business model for Consumables is the same as today: users buy new supplies whenever they need them. Consumables are relatively cheap, which make any business model except the 'durability' model challenging because of the additional costs of the labour needed to make these other models work. A discount on the next set of Consumables for users returning their worn products will help create the volume required for cost-efficient remanufacturing or recycling of the Consumables.

4.2 SARAH'S CIRCULAR BIKE EXPERIENCE
How the above works in the real world is described in a follow-up of Sarah's story in Stage 2, which ended just before she headed out for her trip across the Alps on her bike with rim brakes. The trip turned out to be amazing, with lots of hours spent on famous mountain passes in beautiful surroundings. Together with friends, or alone with her thoughts, she realised that she had made the right decision in refusing to buy a new bike, and instead, enjoying the lightweight bike she already owned. It had been sunny and dry all week, and since there was no need to race down the mountains, she did not miss the disc brakes. Riding in the mountains made Sarah realise once again how fragile the natural systems around her are, and that she too could do more to limit her impact on the environment.

A few years later – we are now in the near future – the inevitable happens. Sarah's bike has reached the end of its technical lifetime. Even though she has done everything she could to maintain and repair her bike, too many parts need to be replaced at once. Because replacement parts are hard to find and expensive, it makes more sense to buy a new bike.

She goes to her local bike shop and is surprised by how she is treated. Instead of being overwhelmed by row upon row of similar looking bikes, half of them with a 'Sale' sign, she enters a shop with displays showing framesets, groupsets and wheelsets. The salesman asks her how he could help her. Sarah explains her experience, goals and wishes for her new bike.

The salesman explains the new approach to finding the right bike for every rider:

1. Start with a bike fit to determine the right bike sizes suiting Sarah's riding style, body measurements and flexibility.
2. Select a Platform consisting of a frame, fork, seatpost, saddle, stem and handlebar, all in line with the results from the bike fit, and in a custom colour scheme if desired.
3. Select a Powertrain bundle based on the estimated number of kilometres per year she will ride and the price she wants to pay. The Powertrain includes a groupset containing shifters, brakes, derailleurs, a chain, cranks including chainrings and a cassette, as well as various options for wheelsets and pedals.
4. Select a Computer, which functions as a data hub with, as a minimum, an option to communicate with the sensors that are integrated in the Platform and the Powertrain to monitor the state of the bike and send this data to the manufacturers. Other options include traditional functions like integration with power meters, heart rate sensors and navigation.
5. Select the Consumables, the fast-wearing and easy-to-replace parts. These include tyres and handlebar tape.
6. Place the order and be patient for a few days for the delivery of her custom-made road bike.

The whole process sounds like the stories Sarah's father told her about how he bought a road bike in the 1970s. The process of being able to hand-pick the frame you like, choose the colour, and select the parts and wheels is very exciting. The prospect of riding something personalised is really attractive.

Two important things are different from the situation in her father's days. First, Sarah will not immediately have to spend the amount of money a bike cost in the past, as she will only buy the Platform and the Consumables. The Powertrain and Computer will remain property of the manufacturer; she will be charged an amount every month, according to the bundle Sarah chose based on her expected annual kilometres. It reminds Sarah of ordering a new mobile phone, which also comes with several options of devices and price plans.

Second, Sarah has the choice between a Platform with the latest technology and pre-used options. The latest technology is cool – the Platforms offered by various brands are all made from new types of biobased fibre composites, with a negative carbon footprint or guaranteed recycled materials. All have performance levels better than anything she has ever seen before. The pre-used Platforms are from older high-end Platform generations that have been completely checked and either repainted by the manufacturer or did not even require new paint because they had only a few scratches. Both the new and the pre-used Platforms come with a lifetime guarantee, even if Sarah were to crash or otherwise damage the Platform.

All of the Powertrain options are a mix of new, reused and refurbished parts, but only on close inspection Sarah finds some signs that a few parts have been used before. The performance of the Powertrain is guaranteed by the manufacturer, and replacement parts and repairs are included in the monthly fee.

Sarah decides to buy a pre-used Platform with a custom paint of her choice, combining this with a medium-level Powertrain on a 3,000-7,500 km/year bundle. The Platform and the Powertrain brands all offer upgrades, such as a new paintjob for the frame and fork after a few years, and a rental option for a lightweight mountain-specific wheelset for a short duration.

The ordering process is simple, and the Platform, Powertrain, Computer and Consumables will be delivered to the shop within 48 hours. The shop will then assemble the bike and make sure that the saddle position matches the bike fit. Sarah pays for the bike fit, Platform and the Consumables in the shop, and signs a contract with the Powertrain and Computer suppliers, who will collect the fee from her bank account on a monthly direct debit. The Platform comes with a guaranteed buy-back price if Sarah wants to change her Platform, so that the Platform manufacturer is able to reuse it. The Powertrain and Computer need to be returned to the manufacturer for reuse at the end of the contract, unless Sarah decides to extend the contract.

A few days later, Sarah picks up the bike. The salesman explains how a bike passport, which can be viewed by scanning the QR code on the frame with her mobile phone, contains all the relevant information related to her bike. A number of sensors in the bike monitor the wear of parts and are able to detect defects that might lead to failure. The data is shared with the Powertrain manufacturer, and Sarah will get instructions on when and how to maintain her bike to make sure it works best. If the sensors detect that certain parts need to be replaced, the passport will send an automated message to the app on her phone, allowing her to book a timeslot at the bike shop, which will receive the required spare part automatically and on time.

Sarah is a happy customer. Her new bike fits better than her old bike, the bike rides amazing, and the custom colour is exactly how Sarah wants it. At the next group ride with her mates, everyone is impressed by the custom paint scheme Sarah selected, and jealous of the low investment in combination with the low-risk monthly subscription for the Powertrain. The environmental benefits of the concept are clear to everyone, and this ignites a conversation about how other industries should follow the example of the cycling industry.

4.3 CIRCULAR DESIGN INCENTIVES

In Stage 2, we identified three negative design incentives in the linear economic model of the cycling industry that resulted in ever shorter product lifecycles – products are designed to Fail, to be Outdated and to be Out-fashioned. To make the new circular business models profitable, these incentives need to be replaced by incentives that will achieve the exact opposite: keeping the materials in use for as long as possible.

1. MAKE IT LAST FOREVER

Durability is the basis of all the parts on a circular bike, no matter whether the rider owns a part or pays for its performance. To be able to make money on a Platform with a lifetime warranty and buy-back guarantee, the manufacturer will have to make a product that will last forever and can be reused again and again.

Pay-for-performance providers have an incentive to make parts last as long as possible, to prevent failure during the contract and to reduce maintenance and repair costs. Being able to reuse or refurbish parts after a contract will be more attractive if the parts have a long remaining lifetime.

2. COMPATIBILITY, ADAPTABILITY & UPGRADABILITY

When deciding on a Platform, customers should not be limited in their choice for a Powertrain because an interface between, for example, a frame and a wheel is not compatible. Design for compatibility, adaptability and upgradability is also essential for Powertrain manufacturers in order to be able to mix new, reused and refurbished components into complete Powertrains.

This incentive to reverse the ever-increasing numbers of fittings offers an enormous opportunity to reduce waste in the cycling industry. It would lead to less unused parts in Boxes at distributors and shops, and it will also make it possible to mix parts from various generations.

3. EASE OF MAINTENANCE & REPAIR

The Powertrain in a pay-for-performance business model requires manufacturers to completely redesign their products for ease of maintenance and repair. To make money in such a business model, labour costs for maintenance and repair need to be minimised to make sure they are covered by the monthly fee paid by the customer.

Materials would ideally 'self-heal' without any intervention, a bit like fluids sealing small holes in tubeless tyres. Sensors communicating with a cycling computer should tell users to clean and lubricate their drivetrain whenever required, so users would look after their product better.

Design for ease of maintenance and repair also aims to minimise a mechanic's 'hands on tool time', by making sure that no time is lost fiddling with parts that have not been designed with disassembly in mind (the hammer will disappear from the workshop) and looking for information or searching for parts.

In the short term, manufacturers should make sure that:
1. products are easy to assemble, disassemble and reassemble in order to clean or replace parts like shifting mechanisms and bottom brackets;
2. maintenance and repair manuals are supplied with the product and are accessible via the bike passport;
3. clear indicators or sensors show when a part is worn and needs replacement;
4. spare parts required for repairs are commonly available and affordable, and preferably delivered automatically to a bike shop that will change the parts.

4. CHOOSING THE RIGHT MATERIALS

Our final design strategy is to select the right material for the function it has to perform, while at the same time cleaning up the supply chain and preventing materials from ending up in a landfill or an incinerator after all. Designers have the ability to create new markets by selecting materials and suppliers that offer alternative materials and production processes. And designers have an enormous influence on the remanufacturing and recycling processes when a product reaches the end of the use phase.

Design using the right materials:
1. use non-toxic materials that are not harmful during the production process, use or post-use phase of a product;
2. use recycled or remanufactured materials instead of new materials;
3. use biobased materials that have been produced responsibly.

Design for remanufacturing and recycling:
1. be clear about which materials are used in the product;
2. tell the end user what to do with the product at the end of the use phase;
3. design a product in such a way that there is enough remaining value in the product for the cycling industry to make the effort to

recover the materials post-use, because they can make money by doing so;

4. make sure that different materials can easily be separated in a recycling process, preferably without specific tools or chemicals; ideally, parts are made from a single type of material.

Involve suppliers:
1. involve suppliers to develop production processes with the lowest environmental impact;
2. oblige suppliers to share Life Cycle Analysis data;
3. oblige suppliers to take back parts for remanufacturing or recycling.

The use of Life Cycle Analysis to compare different materials and manufacturing processes will help to make comparisons based on factual information over the entire lifecycle of a product. It will also provide the information required to develop an Environmental Product Declaration (EPD) for the bicycle as described in Stage 2.

To make the transition to the right type of materials will be a tough journey. There is currently a very limited market for remanufactured, recycled and biobased materials, because of the automatic preference for new materials. This preference is caused by the perceived quality levels, availability and low costs. However, once enough people start asking suppliers questions, and manufacturers start making prototypes with circular materials, a market for reused, refurbished and remanufactured materials, will slowly start to take shape.

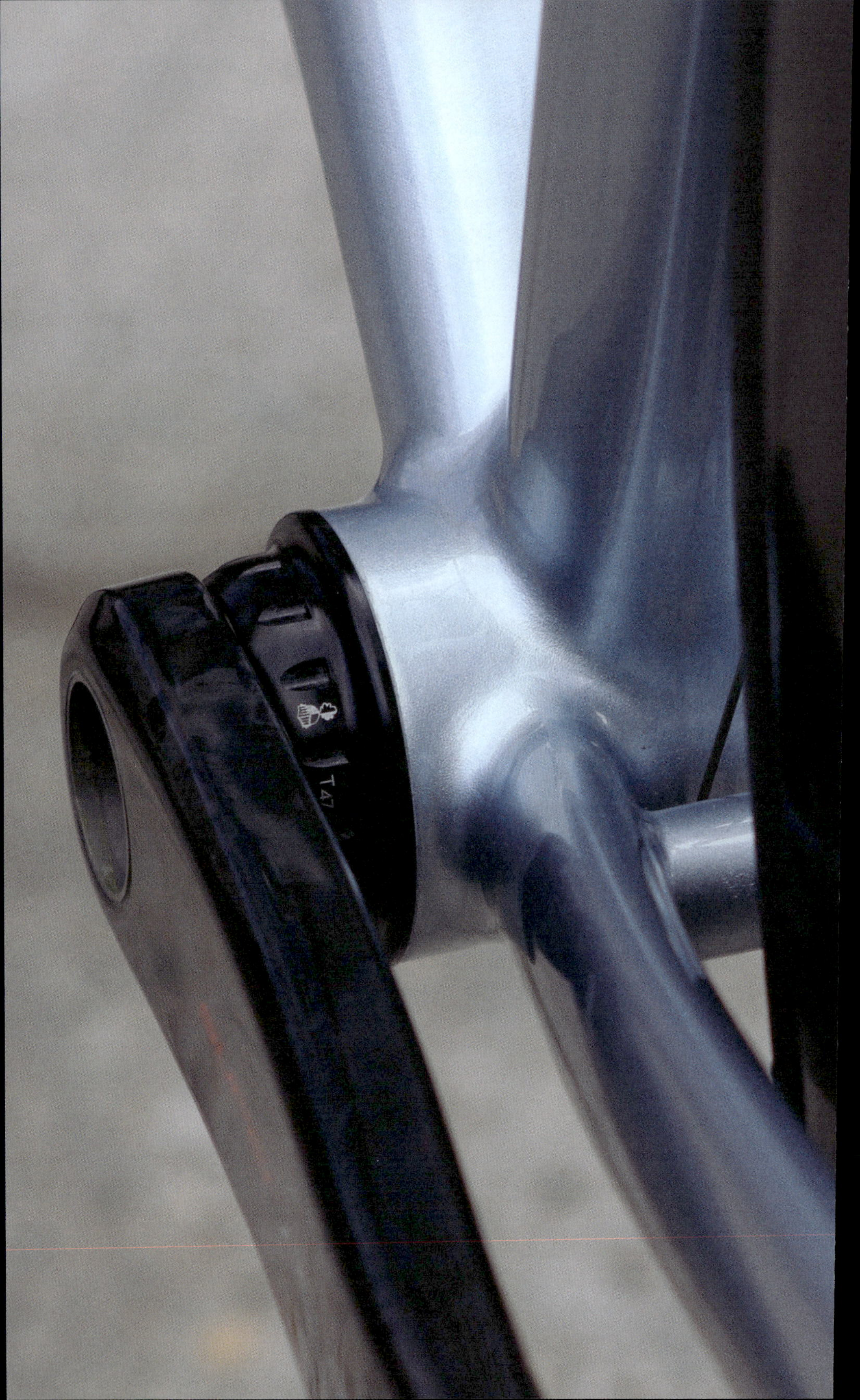
T47

STAGE 5
GETTING THERE

We do not have all the answers to the questions you might be asking yourself by now – no one does. The transition to a circular economy requires changes in the way we think about our current products, about supply chains, about the interaction between manufacturers and users, and the way events are organised. This complex global system cannot be forced to make the transition by a single organisation, a small group of people with the right intentions, or by writing a book. It will require all of the stakeholders in the cycling industry to contribute. What we can do is influence the direction, the speed and the odds that a transition will actually happen.[37]

Transitions are the outcome of an increasing number of small initiatives that slowly but steadily become so large and interconnected that they result in a revolution. The contribution this book wants to make is to help people think differently about existing products and to challenge mindsets. To get individuals and companies to work on initiatives leading to different kinds of designs, to test new business models, and so to learn what works and what does not. To help zoom out and look beyond own positions in the linear economy to what happens in the entire production process. To consider what happens with the products at the end of the use phase. To think about how extended use of a product becomes possible because parts require less maintenance and are easy to maintain, and about making spare parts easy to get and install. To make sure it becomes cool to ride a bike that has been around for years and is well maintained.

When a dream and a deadline come together, an enormous amount of focus and energy can be used for innovation. There were just eight years between President Kennedy's promise to land a person on the moon and get him back safely, and the actual moon landing in 1969. Now, more than 50 years after the first moon landing, we, as a cycling industry working together, should be able to deliver a showcase of a circular bike in eight years, right? How about we set a goal: all riders competing in the road race at the 2028 Los Angeles Olympics ride a circular bike and wear circular kit.

In line with the UCI principles, this bike and kit will be available for everyone to buy in the shop. The stories we tell and write about these products will make this bike the one consumers want to get their hands on. It will signal the take-off of the circular cycling industry and will change the way we think about bike design for ever. For the good of all of us.

5.1 MANAGING THE TRANSITION

The transition to a circular cycling industry will be very similar to previous radical innovations in bike design, only bigger. It requires not just a technical breakthrough, but also a social change to accept the new concept, the development of new supply chains, and the creation of new marketing messages. It is a process that is hard to control, and the complexity will only increase as the number of actors grows.

The introduction of disc brakes on road bikes, for example, required a new design not just for the actual brakes, but also for shifters, forks, framesets, cables, and wheels – the entire system of the bicycle had to change. It began when small brands started to experiment with the adoption of mountain bike brakes for road shifters on cyclocross bikes. Slowly but steadily the design changes on every component reinforced the way the entire system worked, and standards developed, for example for fitting a disc brake calliper to a frame and fork. Bike shops and riders had to get used to the idea of replacing their trusted rim brakes with disc brakes. Once the technology had reached a certain level, and enough riders were convinced that disc brakes really were an improvement, the introduction of the new product really took off. Nowadays almost every new bike design has disc brakes; the transition is complete.

The transition to a circular cycling industry has already started. Phase one of each transition is the 'predevelopment phase' (see Figure 14) in which research and experimentation are carried out at various locations within the industry, without making the headlines in the press and without causing a social change in the larger cycling community. Our UpCycles are an example, as are a number of other initiatives we covered in this book. More initiatives will no doubt be announced before these pages will be printed. It is up to all of us to make the next step, which is when the first circular products will start to enter the market properly in a 'take-off' phase, and the entire system will start to shift towards different designs and business models. In the 'breakthrough' phase, circular products will become the norm: socially, economically and ecologically. During this phase new standards arise and the transition really covers the entire market. Finally, the transition 'stabilises' and is the new normal. Materials used for bicycles and equipment will keep on circulating, waste

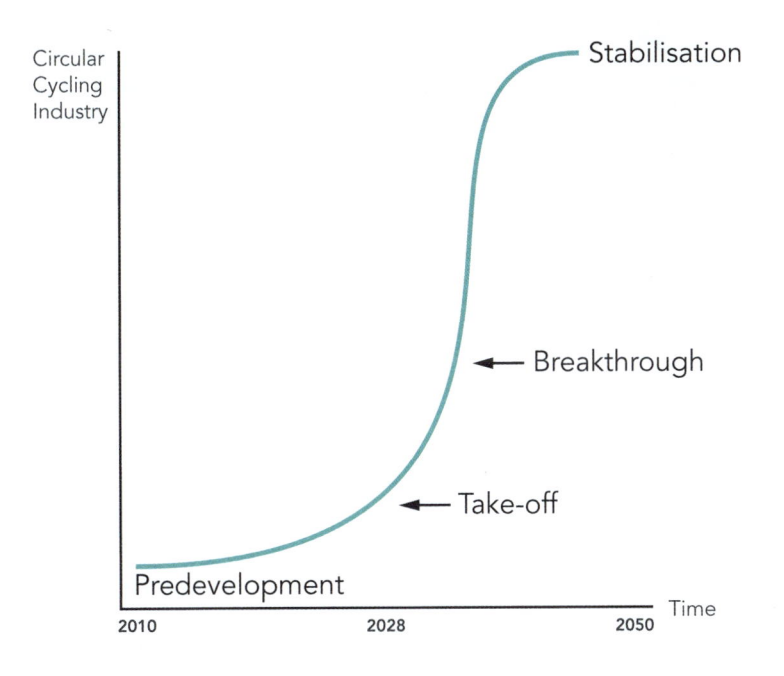

Figure 14: Four phases or a transition.

and pollution will be eliminated, and companies will interact in completely different ways.

To get to a stable circular cycling industry in 2050, all four phases shown in Figure 14 have to be gone through first. The 2028 Olympics are the perfect deadline for the take-off phase. Not only is the timetable manageable, although challenging, but the Olympics are also a perfect stage to show off what has been achieved. Delaying take-off might mean missing the 2050 target, as the breakthrough and stabilisation periods will require several product development generations.

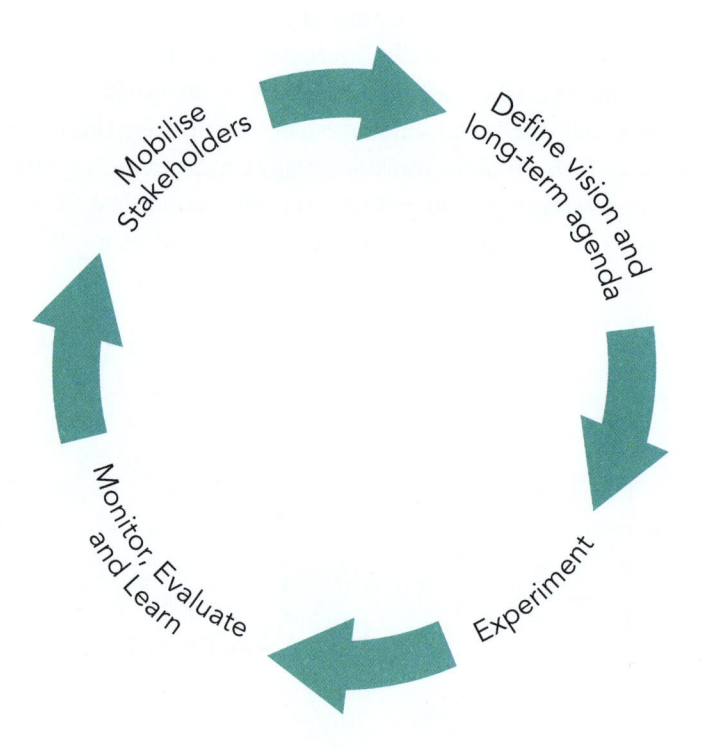

Figure 15: Continuous development cycle for a transition process.

The transition process requires a continuous development cycle of four steps as can be seen in Figure 15, starting with the mobilisation of stakeholders, followed by the development of a long-term agenda, experimentation and evaluation.

JOIN THE SHIFT CYCLING CULTURE FOUNDATION

The international, non-profit Shift Cycling Culture Foundation aims to create awareness, spark conversations and support positive environmental actions, to help materialise the shift to a sustainable cycling community.

Shift Cycling Culture launches campaigns, organises events and sets up pilot projects to inspire brands and the cycling community to start doing things differently. With more care, and less environmental impact. Whether you are part of the industry or just a keen cyclist, check out the projects and see how you can get involved. Shift Cycling Culture can also be used to create awareness about your own project.

Check www.shiftcyclingculture.com and join us!

1. MOBILISE STAKEHOLDERS

There are many different stakeholders in the cycling industry, ranging from large manufacturers and their employees, suppliers and shareholders, to bike shops, race organisers and media. Industry-wide collaboration is necessary to refine this long-term vision and turn it into reality. As we discussed before, we can only escape the tragedy of the commons of the linear economy if we work together, and make a new set of rules fitting a circular economy. Understanding the role of each stakeholder helps to create the required shared vision.

2. DEFINE A VISION AND LONG-TERM AGENDA

The long-term goals for the circular cycling industry are in fact goals for society at large: we all wish to live in a healthy and wealthy society. The European Union has translated this desire into its '100% circular economy by 2050' policy.

What is important now, is to create an inspiring shared vision and long-term agenda for the cycling industry that is aligned with societal aspirations through a process of revolution with a variety of actors in the industry.

This agenda needs to address two important issues:
1. Reduce uncertainty about the direction of the transition, as this creates an impediment to the required system innovation. Uncertainty hinders long-term investments, especially if the innovation interferes with vested interests in the short term.
2. The agenda needs to become part of all the rules and regulations in the cycling sport. If the UCI were to set ambitious goals for bike design, for instance, the procedures to assess new products need to be adjusted to this, potentially allowing products that no-one ever thought about before.

The vision needs to be clear, but cannot be too specific as there will likely be more than a single road towards its realisation. New insights will require constant review and possibly even an adaptation of the final destination.

3. EXPERIMENT

The next step is to create short-term projects with interim goals for stakeholders to work on. These projects create the first products to test in real life. The products can be developed at different levels of the industry – within companies but also in cooperation between companies. Ideally the first pilots benefit the industry as a whole, without causing worries about intellectual property and trade secrets. It is important to select people and organisations with an innovative and cooperative mindset, and the ability to make prototypes that showcase the potential in real life, not just additional paperwork in the form of research projects.

4. MONITOR, EVALUATE AND LEARN

The process described above requires regular evaluation. Just like the preparation of a rider for a large event like the Tour de France, preparation for the transition needs to take into account that the road towards a circular economy will be different from the initial plan. There will be ups and downs along the way and it is important to learn from both.

Another thing to keep in mind, is the process in which parties work together to make the transition work. Has everyone contributed what they signed up for? Are the results influenced by outside trends or vested interests that were not part of the plan? Is there anything that needs to change in the way the involved parties interact? Is the outcome an optimal or perhaps a suboptimal solution that might have a benefit on the short term but blocks the way towards the final goal?

FORMULA 1® NET-ZERO CARBON FOOTPRINT BY 2030

Although it might not seem the obvious sport for serious sustainability targets, Formula 1 has announced a plan to have a net-zero carbon footprint by 2030. The plan includes not just the cars on the track, but all the activities required to make the races happen.[38]

The level of innovation required to create zero-emission cars will challenge the teams like no other regulation in the past. It forces them to develop new drivetrain concepts that eventually will trickle down to ordinary cars – thus having an additional environmental benefit. Offices and other facilities will shift to renewable energy, and as of 2025, Formula 1 events will eliminate single-use plastics, set up incentives for fans to travel to the races using low-carbon alternatives and even improve nature, and create opportunities for local people and businesses.

There are two important similarities between Formula 1 and cycling. They are both highly dependent on the development of high-tech equipment; and both are truly global sports with teams and all their required support faculties continuously travelling around the world.

5.2 CONTRIBUTION OF EACH STAKEHOLDER

As mentioned above, a transition only takes off when enough Stakeholders start contributing to the change. Each Stakeholder has its own role and own interests in the process, which need to be accommodated for. As part of our action plan, we have listed some of the things each stakeholder could do on the short term to realise the take-off as soon as possible.

I. UCI AND NATIONAL FEDERATIONS

Defining the rules of the cycling sport gives the UCI and the National Federations a special place in the transition process. The UCI defines the rules of bike design and both the UCI and the National Federations are responsible for the rules of bike races.

Bicycle Design

As the 'government of the world of cycling', the UCI has a very important role in the development of the road bike. Their rules determine what can and cannot be used in road races, and thereby have a large influence on the bigger brands' innovation agenda.

In Stage 1, we described the philosophy behind the UCI rules. We would like to suggest to the UCI to recognise the importance of taking good care of our planet by adding a few words to their current mission statement:

> *As the summit organisation of world cycle sport, the International Cycling Union (UCI) is the guarantor of the proper application of ethical and sporting regulations.*
>
> *Bicycles shall comply with the spirit and principle of cycling as a sport. The spirit presupposes that cyclists will compete in competitions on an equal footing,* **without harming the environment.** *The principle asserts the primacy* **of planet earth over man**, *of man over machine.*

Including the environment in the UCI philosophy will give investors the required level of assurance that the change towards a circular business model is going to become embedded in the daily operation of the cycling sport. The next step that is required is to translate this philosophy into the actual UCI rules, like the Approval Protocol, which should include rules about the use of hazardous chemicals, pollution and waste. Developing a set of Product Category Rules and requiring brands to include an Environmental Product Declaration as part of the Approval Protocol would make it possible to start measuring the progress towards the goal.

These changes cannot be introduced by the UCI alone, but should be implemented together with manufacturers from the industry. The research and innovation required to comply with the rules could be accelerated by the UCI by funding the development of a cycling-specific Environmental Product Declaration (EPD) and other experiments, provided the results are shared publicly.

Races and clubs
Both the UCI and the National Federations are organisations with contacts with race organisers and cycling clubs. Their pivotal role should be used to provide race organisers and clubs with guidelines in their own language and adapted to local circumstances on how to create a culture amongst riders in which sustainability is the new normal.

The good thing is that many organisers of non-cycling events all over the world face the same challenges and have started reducing their environmental impact. The International Olympic Committee, for example, has published excellent 'Sustainability Essentials' guidelines on their website, which are a great source of inspiration.[39] The Royal Dutch Cycling Federation (KNWU) has a specific page on their website dedicated to sustainability, with initiatives ranging from the reduction of plastic packaging, to stricter rules about riders disposing waste during the races. The KNWU has also teamed up with an energy company to help clubs reduce their energy consumption and switch to energy from renewable sources.[40]

2. THE CYCLING INDUSTRY
As mentioned before, there are already many initiatives in the 'predevelopment stage' of the transition, some of which have been covered in this book. Connecting these initiatives in a joint effort

to develop new materials and business models, makes the 2028 goal achievable. Product generations last for about three to four years before a new version of a frame or groupset is introduced. This gives manufacturers two generations of their product families to transform from linear to circular products. It will be a challenge, but with the right attitude it can be done. To give each stakeholder in the industry a head start, we have come up with a short list of actions that can be done in the short and long term without big investments, alongside existing business.

Platform manufacturers
There are a few things Platform manufactures can start doing immediately. Since they now sell complete bikes, the first step is to make sure that the specifications of each bike are properly documented, accessible and completed with maintenance instructions in a first version of a bike passport. Bike manufacturers have a 'recipe' for every single bike in every single size somewhere in their production and assembly process, including details like the correct length of a chain.

A small pilot to start a buy-back scheme for top-end framesets will help understand how a Quality Control system can be created to check used frames and determine the reuse and refurbishment requirements. It will also result in required design changes for future generations of framesets. A warranty registration system will allow manufacturers to contact buyers of suitable framesets to offer them a buy-back price for the frame.

Manufacturers can then start selling these top-end pre-used Platforms as an alternative to mid-level and low-level Platforms made from new materials. Starting a pilot offering pre-used Platforms directly to consumers in a small and specific market will help to test the best way to market these products, and better understand the customers' requirements for this new product category. By making sure the top-end models stay in the loop through reuse and refurbishment, the average quality of bikes on the road will increase over time, while at the same time less resources are required to make new low- and mid-end Platforms.

The design teams and procurement departments can start searching for alternative materials at their existing suppliers, but also at universities and other R&D-focused institutions, as well as innovative material manufacturers. Getting these materials tested in prototypes as soon as possible, will provide valuable information about possible gaps in terms of performance and/or costs between current and new materials.

Finally, companies can start developing a Platform that makes assembly, repairs and disassembly fast and reliable. Interfaces with parts such as integrated disc brake hoses should become part of the Platform in the form of indestructible brake hoses in the frameset that can be 'clicked' to the brake levers and callipers, without the need to bleed the brake system.

Powertrain manufacturers
Powertrain manufacturers can take a similar approach to product design as Platform manufacturers, with the additional challenge to start testing the pay-for-performance business model.

Just like with the proposed small-scale pilot for the Platform manufacturers, a Powertrain manufacturer can start in a small geographic region with the best conditions to test the concept and gain experience on the best marketing strategy, interaction with users and bike shops, correct price levels, the bike passport, and products returning to the manufacturer. The pilot will lead to an enormous amount of information that can be used to optimise designs, logistics, customer service, etc.

Computer manufacturers
Cycling computers have shown a massive growth in revenue in the last decade. Computers with navigation, multiple sensors and connectivity to smartphones and directly to the internet are providing riders with more information about their rides than ever before.

The development of new types of sensors for monitoring the 'health' of the bike is a market potentially as large as, or even larger than, the rider health data market, because this data will support the Platform and Powertrain manufacturers with the information they need for their product development, maintenance strategies and end-of-life material optimisation.

Developing the option to connect new types of sensors to the existing computers is a simple first step. Getting existing test sensors already used for product development in laboratories out on the road to test the data collection could be the next step. The development of integrated sensors and sharing data with Platform and Powertrain manufacturers via the bike passport will require more time, but can be initiated in the short term without the need for large investments.

What is important for manufacturers is to consider the processing power that these additional sensors require. Software updates will be needed to deal with the new sensors, without jeopardising the performance of older generations of computers, if possible.

Computer manufacturers can also reduce waste significantly by designing the computers for repair and upgrades, and by providing buy-back options for computers that have issues with battery life or processor power.

Consumables manufacturers

Consumables are the products with the shortest lifetime on the bike. This means that the introduction of more sustainable alternatives could reach the highest number of customers in the short term. The challenge for manufacturers is to start using alternative materials as soon as possible.

100% recycled plastic shoe cleats, for example, can probably be produced right now using the same machines the cleats are produced on today, although their colour might not be as perfectly red or black as we are used to. A big bright '100% recycled plastic' sticker on the packaging will surely make customers in more and more markets opt for the recycled version. Selling the screws separately would encourage users to reuse the screws holding their existing cleats in place, instantly saving more materials.

In the long term, more research is needed to develop truly circular products that no longer produce waste or pollution. It will involve choosing alternative materials that as yet might not exist outside laboratories. It will probably also involve developing a system to recover worn tyres through bike shops, so that the carcasses can be remanufactured to new tyres.

Just like for Platform and Powertrain manufacturers, starting conversations with the existing supply chain, universities and research institutes is a good start.

Clothing manufacturers
Clothing manufacturers have not had the attention they deserve in this book, even though this is the part of the industry that potentially is closest to the transition. In the outdoor sports industry, sustainability has been on the agenda for a long time already, and brands without a properly embedded sustainability strategy are bound to lose conscious customers. Brands such as Patagonia and Vaude, which are active in both cycling and other outdoor sports, have been offering sustainable alternatives to linear products for many years, based on a clear long-term strategy.[41]

Without going into too much detail, clothing manufacturers can make considerable steps by improving the durability of products and investigating how alternative materials sourced from renewable (merino wool, for example) and recycled streams can have a positive impact on the environmental footprint of the product.

Another option that is already offered by companies such as Rapha, is a free repair service. Mind you, taking your damaged clothing to a local outdoor or clothing repair shop will also do the trick.[42] Patagonia has extensive care and repair guides on their website and is experimenting with selling reused clothing. Users are asked to return worn clothes to a Patagonia shop for recycling.

For some products, it would be worthwhile to set up experiments with a pay-for-performance model. For instance, consumers could pay per helmet-hour with a guaranteed replacement when the helmet is deteriorated by use or damaged in a crash. This would require clear definitions of when a helmet is considered damaged, so as to prevent users from returning it for the wrong reasons. Apart from taking away the current insecurity about the safety of a used helmet due to age or impact, the helmet would be returned to the manufacturer after the contract expires, to remanufacture or recycle the materials into new helmets.

Bike shops

The transition to a circular economy presents bike shops with numerous opportunities to improve their business models. They can reduce risky stocks of the annual bicycle models and replacement parts. Selling a Platform will allow them to improve service to customers. The assembly of the Platform and Powertrain will once again, be done by the bike shop and the performance contracts of Powertrain manufacturers will generate guaranteed streams of maintenance and repair jobs.

The bike passport will allow them to increase the 'hands on tool time' for mechanics, because they will need less time to figure out which parts are required for repairs. The pay-for-performance concept will force them to improve their internal procedures to make their work as efficient as possible.

As demand increases, bigger bike shops can set up their own large-scale reused and refurbished product line for brands that will not offer these products themselves.

3. EVENTS & RACES

Millions of people all over the world watch one of the many cycling races on a regular basis. Some do this for the thrill and excitement that come from watching the race evolve, others enjoy the Tour de France because of the race's beautiful scenery. Many amateur riders regularly travel to events where they can experience the feeling of riding on these beautiful roads themselves. Rides like the Tour of Flanders attract over 10,000 cyclists the day before the professionals race on the same roads.

These events have a considerable opportunity to diminish their negative effects on the environment. All the materials required to organise the events, the carbon emissions related to the travel of officials, riders and fans, and the waste generated, can be reduced. The good thing is that many organisers of non-cycling events all over the world face the same challenges and have started reducing their environmental impact.

Event organisers

Cycling event organisers are incredibly good at transforming a city into a racetrack. A start/finish area arises, often overnight; kilometres of barriers line the roads with colourful advertisements, and sometimes hundreds of thousands of spectators travel to the city to watch the riders. The day after, everything is removed again and city life returns to normal like nothing happened.

The temporary character of the events already has a substantial circular component – the speed that is required to assemble and later disassemble the racetrack already makes it possible to reuse materials for several events. However, as citizens become more aware of the impact of events, organisers need to do more to reduce the impact of the event on the local community. Especially when it comes to transport emissions, noise, single-use packaging and promotional materials such as flyers that often remain on the streets after the event has moved on.

Event organisers can make a switch to low-noise and zero-emission hydrogen and electric vehicles over the next few years when they contract new mobility providers. Power for temporary facilities should come from renewable sources, for example by replacing diesel power generators by hydrogen versions. Spectators should be incentivised to travel to the race by bike or public transport. Using less single-use materials and better recycling facilities can have a large effect on the production of waste. The eco-sustainable recycling project launched by race organiser RCS Sport as part of their Ride Green project collected 75,758kg of litter during the 2018 Giro d'Italia, of which 90% was recycled.[43]

Instagram

4,555 likes

lachlanmorton Huge year, best one yet. Spain 🛫 Albania 🚴 Kosovo 🚴 Macedonia 🚴 Greece 🚴 Bulgaria 🚴 turkey (outskirts) 🛫 spain 🛫 Australia (Nats,TDU,Cadel's, suntour) 🛫 spain 🚌 France (haut var) 🚌 spain 🛫 Italy (Strada) 🛫 spain 🛫 Italy (Coppi) 🛫 California(Death Valley 🛫 COLORADO 🛫 Australia (em and ham wedding) 🛫 COLORADO 🛫 California (tour of Cali) 🛫 Kansas City 🛫 COLORADO 🛫 emporia (dirty Kanza) 🛫 Sweden 🛫 spain 🚌 France(mt ventoux) 🚌 spain 🛫 UK (gb duro) 🛫 spain 🛫 Vietnam 🛫 COLORADO (Leadville, Utah, CO trail) 🛫 San Fran 🛫 COLORADO 🛫 UK (three peaks) 🛫 spain 🛫 Italy (One days) 🛫 spain 🛫 Japan (Japan cup) .... 😴 going into hibernation, need to rest and absorb everything that's happened so I can prepare for an even bigger 2020. Photo @ashleygruber

---

A PRO RIDER'S 2019 CARBON FOOTPRINT: 23,720KG

Most of a pro cyclist's life is spent on a bike or travelling to the next training camp or race. On 21 October 2019, EF-Education First pro rider @LachlanMorton gave insight into his 2019 journeys on Instagram.[44] It is an impressive schedule with at least 29 trips.

We had a go at calculating Morton's 2019 carbon footprint using this schedule, assuming he travelled to and from the closest airports to his destinations, and that the logistics experts on his team had managed to find direct flights. The CO_2 emissions we calculated include his flights only, i.e. things such as car travel, heating and air conditioning of his accommodation, and food preparation are not included. We entered the airport of origin and the destination airport in the online CO_2 compensation calculation tool provided by NGO Trees for All to calculate the total carbon footprint.[45]

The result was stunning: 23,720kg of invisible climate-disrupting gases, equal to the yearly emissions caused by about five average human beings.

Sustainability should become part of the communication strategy of every event organiser. Talking about sustainability at events will help make it an issue in the community and change the culture.

Teams and riders

As one of the most attractive TV sports, professional cycling reaches many people. This comes with a responsibility for teams as well as riders. Just like in the life of an ordinary cyclist, waste that is created to make riders perform as good as possible cannot easily be avoided. Pro teams have a large fleet of vehicles travelling to the races, riders and staff regularly fly to training camps and races, gear wears and breaks, supplies come in disposable packaging, etc.

A number of teams recognised their ecological impact and started programs to make the team and their partners more aware, as well as a CO_2 compensation program to invest in the development of new forests. Team Sunweb has been an ambassador of the NGO Trees for All since 2010.[46] Together with Trees for All, the team calculates the annual carbon footprint caused by their facilities, vehicles and flights. The carbon emissions are then compensated by Trees for All by investing in forest conservation and plantation projects in several countries. In 2020, Team Sunweb will invest specifically in the development of a forest close to their training facilities in Limburg, the Netherlands. Team Deceuninck-Quickstep launched #itstartswithus at their 2020 team presentation.[47] Just like Team Sunweb, it calculates and compensates their carbon emissions by planting trees, in Uganda and close to Mont Ventoux in France. The program also includes a 'manifesto of changes' with measures to reduce the use of plastics, educate staff and riders, and raise awareness among partners and suppliers.

Together with the UCI and race organisers, the teams can switch to hydrogen cars, buses and trucks in the next few years. Hydrogen-fuelled vehicles could replace their current fleet without the range restrictions that come with electric vehicles. With more hydrogen fuel stations opening every month all over Europe, and the possibility to supply the entire fleet via a mobile hydrogen fuel station, the colourful caravan that travels across the continent could become an example of the zero-emission option in transportation.

The transparency of and investments by the teams send a clear message to their suppliers, other teams and the general public: the issue of sustainability is something that affects everyone. It signals that being the best possible rider includes being a conscious rider, aware of his/her impact and the possibilities to contribute to positive change.

4. MEDIA

The media have an important role in telling the story of the introduction of new products that will be developed over the next few years. They too can help to accelerate the transition, by simply asking manufacturers about the steps they are taking to create products that fit in a circular economy. Writing stories about the necessity of the transition, as well as about new developments, will inspire both businesses and consumers. Media have the ability to create the required public support by favouring and praising brands that try out new designs, materials and business models. Accepting that failure is part of the journey, will hopefully encourage journalists to think beyond a clickbait title such as 'eco-friendly option not as good as standard option' and write honest stories about both difficult and exciting developments.

A special section in a magazine or website that publishes stories about these developments would give conscious consumers a place where they can learn more about the developments that are taking place in the current 'pre-development' phase, and could serve as a source of inspiration for designers across the industry.

5. CONSUMERS

We all seem to have them. A favourite thermal, pair of shoes or pair of sunglasses. You use it way more often than the others you have in your closet. Often your favourites were more expensive than the stuff you bought on impulse because it was on sale. When a favourite finally breaks down, you're heartbroken, and you can spend hours looking for the exact same product – often without success. What if every piece of cycling equipment you owned would be such a favourite? Your closet would be a lot emptier, and you would save money by not buying stuff you hardly use.

There are at least five things you can do today while manufacturers develop their circular products:

1. **Buy less.** If a part or piece of kit is still in good order, keep on using it! Don't go with the latest marketing trend.

2. **Buy better.** Buy one super nice shirt that will last forever, instead of three cheap shirts that you will only use a few times because they are not so nice after all.

3. **Maintain, maintain, maintain!** Spend more time cleaning and maintaining your bike. If you do not know how, or do not have the time, let your bike shop do it for you.

4. **Repair instead of replace.** Having parts or clothes repaired is often cheaper and saves a lot of waste.

5. **Ask your bike shop and cycling buddies about sustainable alternatives.** If enough people start asking questions, shops and manufacturers will get the message that there is a demand, and will start to invest in the required innovation process.

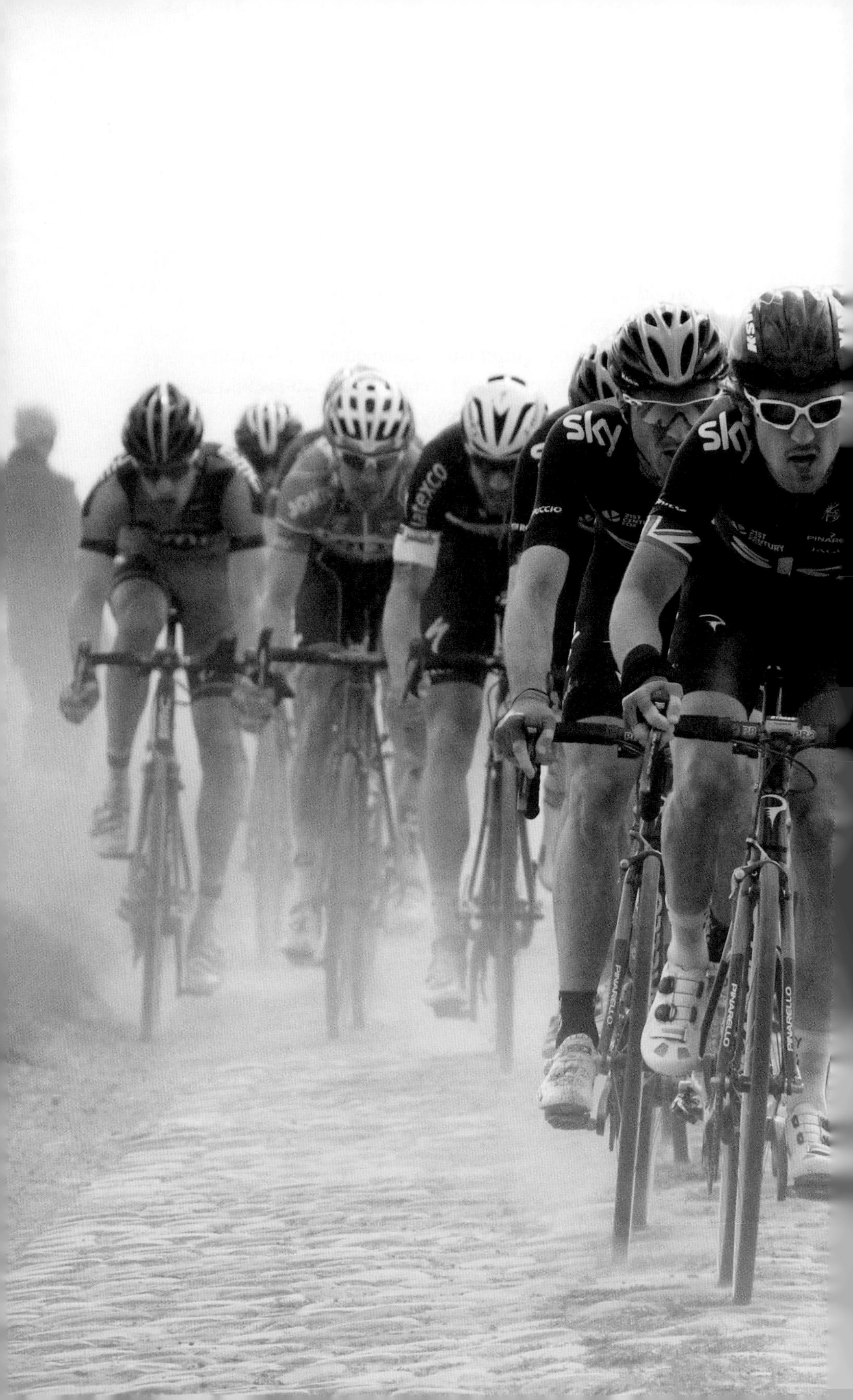
sky
sky
latexco
PINARELLO
PINARELLO

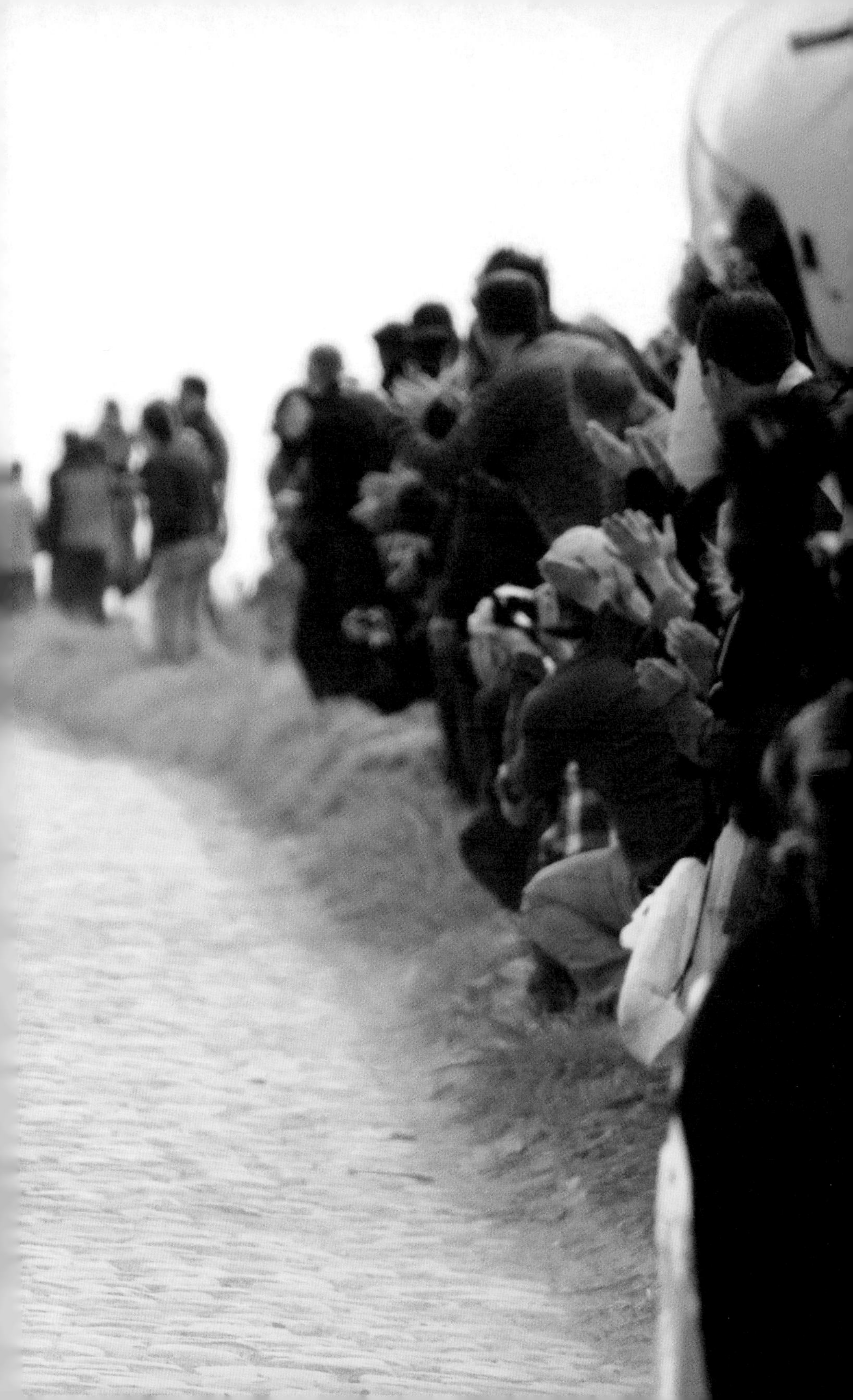

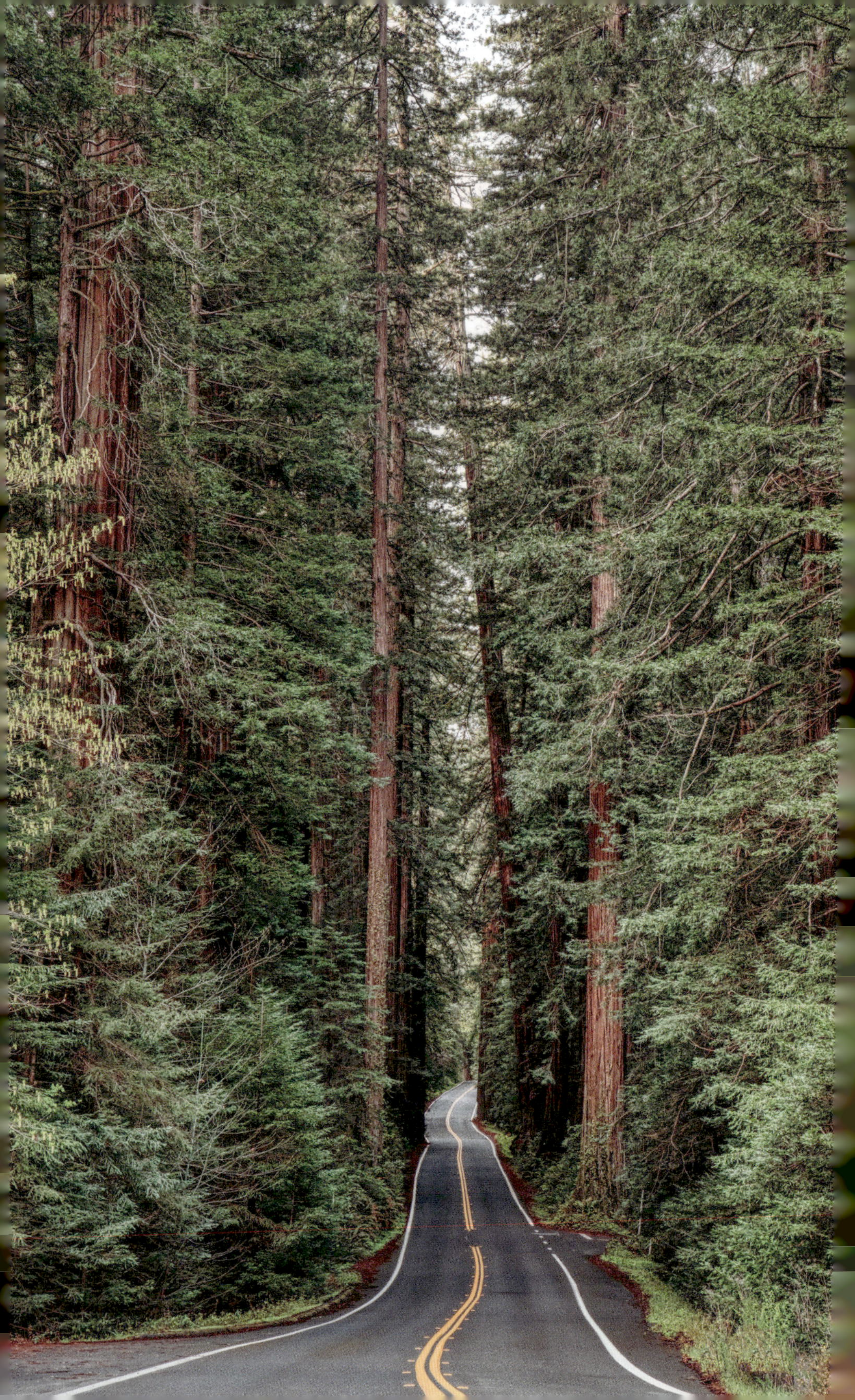

FINISH
STAND UP FOR A REVOLUTION

Human psychology is probably the most important thing that keeps us from making a fast transition to the circular economy. The idea that your own actions are insignificant and will not lead to the required change is not exclusive to you – most people experience exactly the same. Nobody is perfect, nor does our current linear economic system allow us to live a normal life without waste and pollution.

To get to a circular cycling economy, we need much more than marginal gains, we need a revolution.

YOU CAN MAKE A DIFFERENCE

The truth is, if everyone keeps thinking he or she cannot make a difference, nothing will happen. People have an immense power to create positive change by standing up and making a start, however small. It will also help you feel better, because you are actually doing something. It works similarly to starting with a new training program for a big cycling event after a long winter with little time on your bike: it is tough in the beginning, because your body is no longer used to it, but it feels better every time you go out for a ride.

What can you do yourself? Talk to your friends and family about your wish to make the transition a reality, and inspire them to follow your actions and make a start themselves. Dive into your shed to check which parts in your Box you can actually reuse yourself or pass on to someone who can. Start a conversation with your design- or sales-team and get inspired by other industries or something that you read in, for example, the newspaper. Ask your suppliers about their ideas and solutions. You will discover that you are not the only one who wants to contribute, and even that many of the people you talk to are already contributing.

Next time you have to decide whether to buy or repair, visit your local bike shop for a repair. If you have to buy, start a conversation with the shop about the sustainability of the product you plan to buy. Make a conscious decision and go for an alternative with less ecological impact compared to the product you have always been using.

Yes, just like in a stage race where every rider faces challenges, you will have setbacks. Some people will disagree with you, or the ecological alternative is slightly more expensive. But if you are able to inspire, say, five people a year, and they start doing the same thing, the potential is huge.

RETAINING THE VALUE OF THIS BOOK

You have reached a point where this book is about to enter the post-use phase. Having read these pages, printed on FSC-certified recycled paper, you now have the choice between a linear and a circular end-of-life scenario. You have the option to throw this book into the bin, because you thought it was worthless, or because you do not have the opportunity to recycle paper. You can also place the book on a shelf in your bookcase, which is pretty much the same as storing old bike parts in a Box in your shed.

You can also increase the use of this book by referring to it when you start the development of circular business models or designs. This would not only increase the duration of the use phase of the book, it would also increase the value of the book as the ideas in it become a reality. Once you are done using this book, you can pass it on to someone you think might enjoy reading it. The book would be reused, hopefully more than once – retaining its value.

It sounds crazy from a linear economic perspective, but we would rather have this copy of the book passed around than to sell more of them. The reason? We did not publish this book to make a fortune – we published it to spark conversations and accelerate the transition to a circular economy.

Of course, the information in this book has only temporary value. In a few years' time, circular business models and designs will become the new norm, and the value of this book will be reduced to a good memory of 'how things used to be back in the 2020s', or a few cents for a paper recycler. Whatever you do with it, please make sure you dispose of this book properly to make sure that it will not end up in nature or on a landfill.

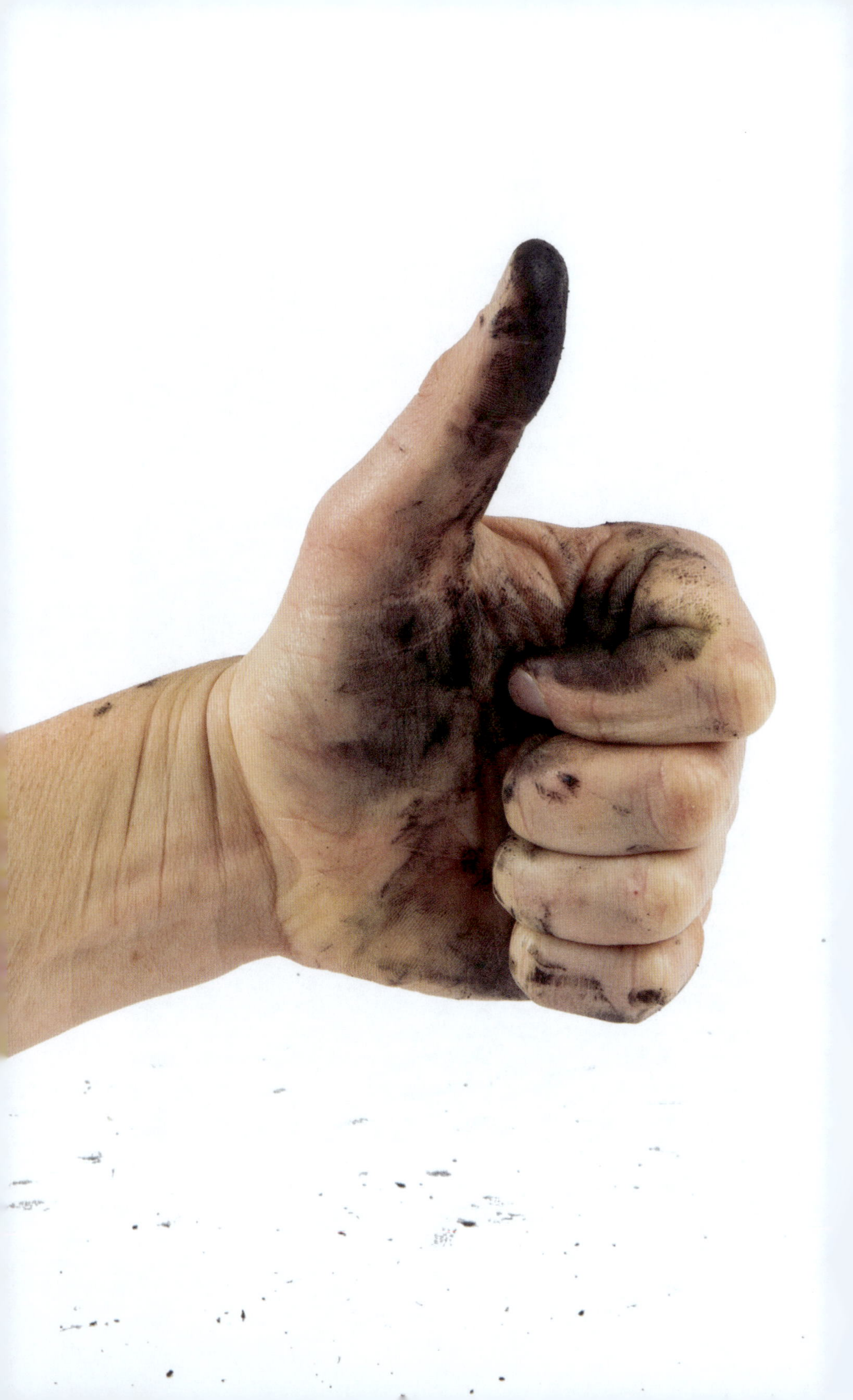

THANKS!

To Eveline and Floortje, who gave us the opportunity to spend so much time working on realising our ideas since the Circular Cycling journey started – away from home or deep in thoughts at the dinner table.

To our team that made the book a reality during strange times: Wardy Poelstra, Erwin Postma, Erik Diekstra, Lucas Reinds and Robert-Jan van Noort. Wim Bronsvoort for his countless revisions, and Sjors Kurvers, whose inside knowledge of the cycling industry proved very valuable.

To all the people inside and outside the cycling industry who somehow contributed to Circular Cycling and share our goal to make the industry sustainable: Debbie Appleton, Samuel Beltman, Ken Bloomer, Saskia Boer, Joost Brinkman, Vince Crossley, Ruud van Dam, Jane Dennyson, Fiola Foley, Chretien Herben, John van Herwerden, Jeroen Hinfelaar, Jeroen van den Hout, Erik van der Hout, Ingo Jansen, Troy Jones, Wouter Keijzer, Lian van Leeuwen, Mantijn van Leeuwen, Marianne van Leeuwen, Marc Lensen, Ruben van Loon, Gerrit Middag, Elise de Reyer, Sami Sauri, Menno Smeelen, Geeske Tamsiran, Bram Tankink, Hein van Tuijl, Victor Vijfwinkel, Margo de Vries, Staffan Widell, Alex Weller, Lennard van Winden.

ABOUT THE AUTHORS

ERIK BRONSVOORT, 1981

Erik bought his first mountain bike, a Gary Fisher Big Sur, in 1997 and was hooked to cycling immediately. He studied Civil Engineering at Delft University of Technology, joined the student cycling club full of like-minded bike nerds, and worked in a bike shop on Saturdays to be able to buy new bike parts to tune his bikes. His modest racing career peaked when he finished the mountain bike Transalp Challenge in 2004.

As an engineer, Erik worked on track replacement in the London Underground for a few years, before becoming an innovation project manager at a large construction company in the Netherlands. Since 2009, he has been working on sustainability projects, first as an employee, then as a self-employed project manager supporting both corporates and start-ups with his project management skills and knowledge about technology and sustainability.

Founding Circular Cycling in 2018 proved to be the best way to learn about how a circular economy could really work, and Erik now uses this knowledge to train companies in the development of circular business models and manage their implementation.

MATTHIJS GERRITS, 1982

Matthijs' first real bike was a Cannondale M900 that he fully customised in the 1990s. When Matthijs founded Circular Cycling, his shed was the best source for parts. His passion for (bike) technology made him start a study in Industrial Design at Delft University of Technology, before he switched to Leiden University to study History. Like Erik, Matthijs worked at a bike shop and joined the student cycling club.

Matthijs was lured back into the cycling industry by a large distributor at the time when IT systems became an important factor in the success or failure of brands. As an IT manager, Matthijs thought out, implemented and integrated a wide variety of front-end and back-end systems, from web shops and API connections, to Warehouse Management and Product Information Systems. He was also an active participant in DST, an online database for digital product information exchange within the cycling industry.

He currently works for NIBE, an international organisation that provides tools and consultancy for Life Cycle Analysis and Environmental Product Declarations. At NIBE, Matthijs combines his in-depth circular economy knowledge with his IT expertise.

'IN 2050, WE LIVE WELL, WITHIN THE PLANET'S ECOLOGICAL LIMITS. OUR PROSPERITY AND HEALTHY ENVIRONMENT STEM FROM AN INNOVATIVE, CIRCULAR ECONOMY WHERE NOTHING IS WASTED AND WHERE NATURAL RESOURCES ARE MANAGED SUSTAINABLY, AND BIODIVERSITY IS PROTECTED, VALUED AND RESTORED IN WAYS THAT ENHANCE OUR SOCIETY'S RESILIENCE. OUR LOW-CARBON GROWTH HAS LONG BEEN DECOUPLED FROM RESOURCE USE, SETTING THE PACE FOR A SAFE AND SUSTAINABLE GLOBAL SOCIETY.'

'BICYCLES SHALL COMPLY WITH THE SPIRIT AND PRINCIPLE OF CYCLING AS A SPORT. THE SPIRIT PRESUPPOSES THAT CYCLISTS WILL COMPETE IN COMPETITIONS ON AN EQUAL FOOTING, WITHOUT HARMING THE ENVIRONMENT. THE PRINCIPLE ASSERTS THE PRIMACY OF PLANET EARTH OVER MAN, OF MAN OVER MACHINE.'

circular•cycling

SHARE YOUR THOUGHTS

We would love to hear your thoughts, questions and ideas on this subject. Whether it's at a personal level or more system-focussed, what steps can we start taking to contribute to a revolution? Please join the conversation on our social media pages.

And if you're feeling creative, feel free to sketch out your ideas below and share them.

in 📷

Share a photo or thought using:
#circularcycling @shiftcyclingculture

Shift Cycling Culture

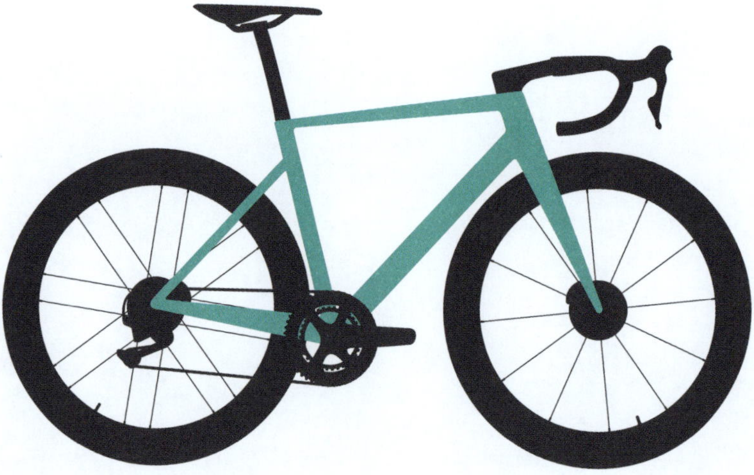

'IT WOULD BE A WIN-WIN-WIN SITUATION. **CUSTOMERS** WIN BECAUSE THEY ARE ABLE TO RIDE MORE RELIABLE PRODUCTS THAT REQUIRE LESS MAINTENANCE. **THE INDUSTRY** WINS BECAUSE IT CAN CREATE EQUAL OR MORE VALUE WITH LESS MATERIAL. FINALLY, **THE PLANET** WINS BECAUSE THERE WILL NO LONGER BE ANY WASTE OR POLLUTION, AND NATURE WILL GET A CHANCE TO REGENERATE.'

circular·cycling

NOTES

Prologue

1. United Nations, 1992: *United Nations Framework Convention on Climate Change*. https://unfccc.int/files/essential_background/background_publications_htmlpdf/application/pdf/conveng.pdf
2. World Bank, *Poverty headcount ratio at $1.90 a day (2011 PPP) (% of population)*, 1987-2018. https://data.worldbank.org/topic/poverty?end=2018&start=1987
3. United Nations, Department of Economic and Social Affairs, *World: total population*. https://population.un.org/wpp/Graphs/DemographicProfiles/Line/900
4. 'Briefing: Global warming 101. The past, present and future of climate change', *The Economist*, Sep 21 2019. https://www.economist.com/briefing/2019/09/21/the-past-present-and-future-of-climate-change?cid1=cust/climateissue/n/bl/n/20191230n/owned/n/n/climateissue/n/n/EU/369848/n
5. M. Murphy, 'Technology Quarterly: China has never mastered internal-combustion engines. Electric cars will be different', *The Economist*, Jan 2 2020. https://www.economist.com/technology-quarterly/2020/01/02/china-has-never-mastered-internal-combustion-engines
6. European Commission, *Environment Action Programme to 2020*. https://ec.europa.eu/environment/action-programme

Stage 1: the complex world of cycling and sustainability

7. This norm was revised in 2005 (EN ISO 14781:2005) and merged into a broader norm for bicycles sec in 2014 (EN ISO 4210-2:2014), which again was revised in 2015 (EN ISO 4210-2:2015). The (English) draft text of the original norm from 2004 can be accessed for free online on the Dutch NEN website: NEN, 2003: *Racing bicycles - safety requirements and test methods. Ontwerp NEN-EN 14781*. www.nen.nl/pdfpreview/preview_44926.pdf
8. UCI, (not dated): *Approval Protocol for Frames and Forks*. www.uci.org/inside-uci/constitutions-regulations/equipment
9. G. Hardin, 'The Tragedy of the Commons', *Science* 162 (1968) 1243-1248.

Stage 2: the linear cycling economy

10. E. Achterberg, J. Hinfelaar en N. Bocken, 2016: *Master Circular Business with the Value Hill*. https://assets.website-files.com/5d26d80e8836af2d12ed1269/5dea-74fe88e8a5c63e2c7121_finance-white-paper-20160923.pdf
11. Markus Krajewsk, 'The Great Lightbulb Conspiracy', *IEEE Spectrum*, Sep 24 2014. https://spectrum.ieee.org/tech-history/dawn-of-electronics/the-great-lightbulb-conspiracy

12. World Economic Forum, 2018: *The Global Risks Report 2018, 13th Edition.*
    http://reports.weforum.org/global-risks-2018/grim-reaping
13. Carbon Tracker Initiative: *Climate change can only be tackled by reducing emissions. Narrative infographic,* January 2020.
    www.carbontracker.org/wp-content/uploads/2020/02/CTI_Narrative_Infographic_Jan_2020_3_NE.png?mc_cid=cb4ecf3f18&mc_eid=b98dc419c8
14. CBS, 2020-21-2: *De Nederlandse economie - Circulaire economie in Nederland.*
    www.cbs.nl/nl-nl/achtergrond/2020/08/circulaire-economie-in-nederland
15. CIRCLE Economy: *The Circularity Gap Report 2020.*
    www.circularity-gap.world/2020
16. L. Peake, C. Brandmayr and B. Klein, 2018: *Completing the circle: creating effective UK markets for recovered resources.*
    www.green-alliance.org.uk/completing_the_circle.php
17. www.grondstoffenscanner.nl/
18. R. Bezemer, 'Rubberbedrijven staan in de rij voor Russische paardenbloem', *Kunststof en Rubber* Jul 2017.
    www.kunststofenrubber.nl/download/Russische%20paardenbloem%20Vredestein-WUR.pdf
19. EPEA: *Energy Balance for the Recycling of Butyl Rubber Bicycle Inner Tubes.*
    www.schwalbe.com/schwalbe_recycling/nl/Energiebilanz_EN_20190625.pdf
20. www.ghgprotocol.org
21. R. Johnson, A. Kodama and R. Willensky, 2014: *The Complete Impact of Bicycle Use. Analizing the Environmental Impact and Initiative of the Bicycle Industry* April 2014.
    https://dukespace.lib.duke.edu/dspace/bitstream/handle/10161/8483/Duke_MP_Published.pdf?sequence=1&isAllowed=y
22. The electricity use of the average Dutch household is 2830 kWh.
    https://www.milieucentraal.nl/energie-besparen/snel-besparen/grip-op-je-energierekening/gemiddeld-energieverbruik/
23. www.environdec.com

Stage 3: creating value in the circular model

24. www.ellenmacarthurfoundation.org
25. www.circle-economy.com
26. C. Bakker et al, 2014: *Products that Last. Productontwerpen voor circulaire businessmodellen.*
27. www.circonl.nl/international/
28. ELG Carbon Fibre Ltd, 2017: *LCA benefits of rCF (Conference presentation for Composite Recycling & LCA,* Stuttgart March 9 2017).
    www.elgcf.com/assets/documents/ELGCF-Presentation-Composite-Recycling-LCA-March2017.pdf
29. www.specialized.com/us/en/carbon-fiber-recycling-program
30. T. Adams, 'Fritz Vollrath: "Who wouldn't want to work with spiders?"', *The Guardian online,* Jan 12 2013.
    www.theguardian.com/science/2013/jan/12/fritz-vollrath-spiders-tim-adams

[31] J.-F. Bastin et al, 'The global tree restoration potential', *Science* 365 (2019) 76-79. https://science.sciencemag.org/content/365/6448/76
[32] https://limburgbike.brightlands.com

Stage 4: action plan for a circular cycling industry
[33] www.cannondale.com/en/app
[34] www.ptc.com/en/case-studies/cannondale
[35] www.sram.com/en/life/stories/introducing-sram-axs-web
[36] www.stichtingdst.nl/en/platform-dst/

Stage 5: getting there
[37] R. Kemp en D. Loorbach, 2003: *Governance for sustainability through transition management.*
www.researchgate.net/publication/2883708_Governance_for_Sustainability_Through_Transition_Management
[38] https://corp.formula1.com/corporate-responsibility/
[39] www.olympic.org/sustainability-essentials
[40] www.knwu.nl/duurzaamheid
[41] https://eu.patagonia.com/nl/en/the-activist-company.html and https://csr-report.vaude.com
[42] www.rapha.cc/nl/en/repair-service
[43] www.giroditalia.it/eng/ridegreen/
[44] www.instagram.com/p/B33boRtBEyv/ @lachlanmorton
[45] https://treesforall.nl/uitstoot-berekenen/
[46] https://teamsunweb.com/about-us/
[47] www.deceuninck-quickstep.com/en/team/about/itstartswithus

All URLs were checked in April 2020.

PHOTO CREDITS

Cover	Tim Bardsley-Smith, TBSphotography
2	Patrick Langwallner on Unsplash
6	Tim Bardsley-Smith, TBSphotography
8	Unkown photographer, wikipedia
16	Simon Connellan on Unsplash
22	Cor Vos, Fotopersburo Cor Vos
28	Unkown photographer, Google
30/31	UCI Approval Protocol, UCI.org
37	Cameron Venti on Unsplash
38	Dominik Vanyi on Unsplash
48	Aaron Burden on Unsplash
53	Saskia Boer, saskiaboer.com
54	Andrew Gook on Unsplash
65	The authors
66/67	Saskia Boer, saskiaboer.com
76	Casey Horner on Unsplash
81	Marie Westphal on Unsplash
86	Warren Wong on Unsplash
88/89	Tim Bardsley-Smith, TBSphotography
98	The authors
108	Coen van den Broek on Unsplash
116	Alex de Kraker, St Joris Cycles
136/137	Simon Connellan on Unsplash
138	Cameron Venti on Unsplash
142	Saskia Boer, saskiaboer.com
144/145	The authors
156/157	Cameron Venti on Unsplash

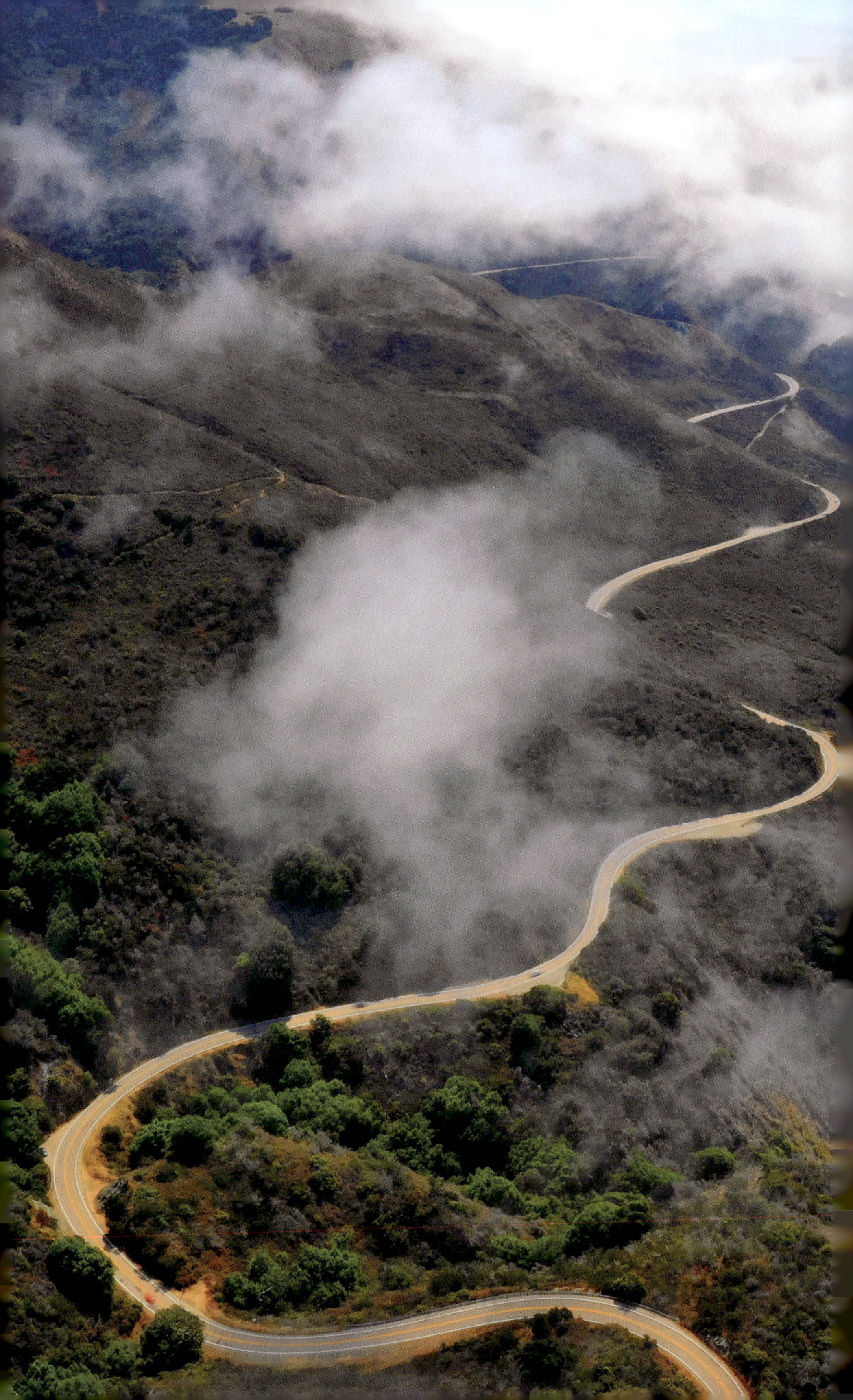

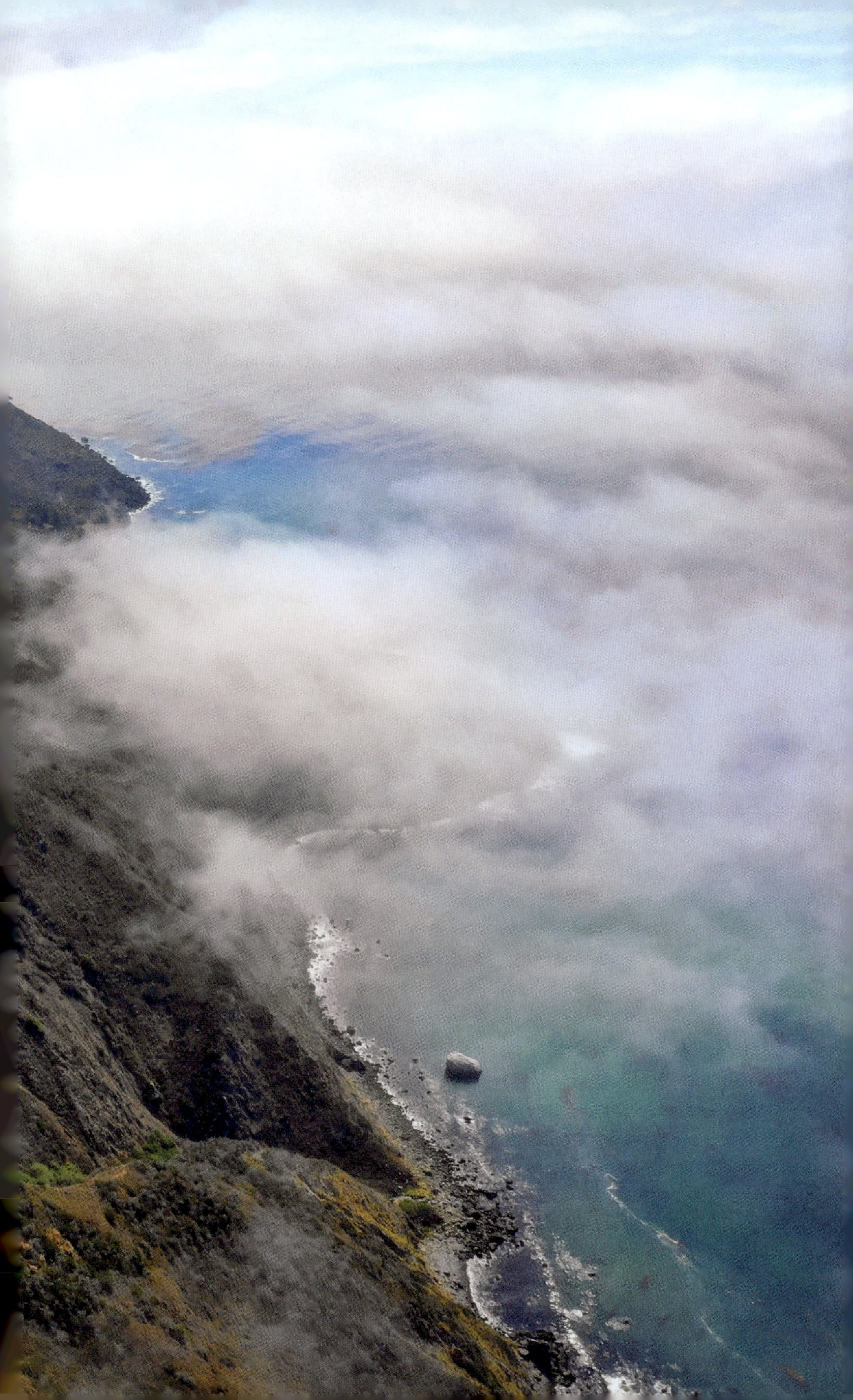

CIRCULARCYCLING.COM